Worksheets

For Classroom or Lab Practice

Jeanette Shotwell

Central Texas College

Beginning Algebra

FIFTH EDITION

Elayn Martin-Gay

PEARSON

Prentice Hall

Upper Saddle River, NJ 07458

Editorial Director: Christine Hoag
Editor-in-Chief: Paul Murphy
Sponsoring Editor: Mary Beckwith
Assistant Editor: Georgina Brown
Senior Managing Editor: Linda Mihatov Behrens
Associate Managing Editor: Bayani Mendoza de Leon
Project Manager, Production: Robert Merenoff
Art Director: Heather Scott
Supplement Cover Manager: Paul Gourhan
Supplement Cover Designer: Victoria Colotta
Operations Specialist: Ilene Kahn
Senior Operations Supervisor: Diane Peirano
Mgr. Visual Research and Permissions: Karen Sanatar

© 2009 Pearson Education, Inc.
Pearson Prentice Hall
Pearson Education, Inc.
Upper Saddle River, NJ 07458

Pearson Prentice Hall™ is a trademark of Pearson Education, Inc.

The author and publisher of this book have used their best efforts in preparing this book. These efforts include the development, research, and testing of the theories and programs to determine their effectiveness. The author and publisher make no warranty of any kind, expressed or implied, with regard to these programs or the documentation contained in this book. The author and publisher shall not be liable in any event for incidental or consequential damages in connection with, or arising out of, the furnishing, performance, or use of these programs.

Printed in the United States of America

10 9 8 7 6 5 4 3 2 1

ISBN-13:   978-0-13-603114-7   Standalone
ISBN-10:   0-13-603114-5   Standalone
ISBN-13:   978-0-13-603080-5   Component
ISBN-10:   0-13-603080-7   Component

Pearson Education Ltd., London
Pearson Education Singapore, Pte. Ltd.
Pearson Education Canada, Inc.
Pearson Education—Japan
Pearson Education Australia PTY, Limited
Pearson Education North Asia, Ltd., Hong Kong
Pearson Educación de Mexico, S.A. de C.V.
Pearson Education Malaysia, Pte. Ltd.
Pearson Education Upper Saddle River, New Jersey

Worksheets for Classroom or Lab Practice

Beginning Algebra, Fifth Edition

Table of Contents

Chapter 1 Review of Real Numbers
Section 1.2 Symbols and Sets of Numbers

Learning Objectives
1. Use a number line to order numbers.
2. Translate sentences into mathematical statements.
3. Identify natural numbers, whole numbers, integers, rational numbers, irrational numbers, and real numbers.
4. Find the absolute value of a real number.

Vocabulary.
Use the choices to complete each statement.

Absolute value	Natural numbers	Whole numbers
Inequality	Rational numbers	Zero
Integers		

1. _____ is neither positive nor negative.

2. The _____ are the entire set of natural numbers and 0.

3. _____ are numbers that cannot be expressed as a quotient of two numbers.

4. The distance a real number c, is from zero is called the _____. It is denoted by

     _____ .

5. The symbols $\neq$, $\leq$, and $>$ are called _____ symbols.

6. The _____ are all the numbers in the set $\{1, 2, 3, 4,...\}$

7. The _____ are the set of numbers that can be written as a quotient of two numbers.

8. The set of _____ are all the numbers that can be represented on a number line.

9. The _____ are the set of all the positive and negative whole numbers.

Objective 1
Insert $<$, $>$, or $=$ in the appropriate space to make the statement true.

10. 15 ___ 12

11. 7 _____ -10

12. -3 _____ -2

13. 1.24 _____ -2.24

10._____

11. _____

12. _____

13. _____

Are the following statements true or false?
14. $11 \leq 11$

15. $3(2+4) < 2(3+4)$

16. $24 > 34$

17. $-1.78 \leq -1.68$

14. _____

15. _____

16. _____

17. _____

Objective 2

Write each sentence as a mathematical statement.

18. Three is less than or equal to six.

18. _____

19. Negative eight is greater than negative fifteen.

19. _____

20. Seven is not equal to negative seven.

20. _____

21. Five is greater than or equal to four.

21. _____

Objective 3

Tell which set(s) of numbers each of the following belong to: natural numbers, whole numbers, integers, rational numbers, irrational numbers, and real numbers.

22. 5

22. _____

23. -8

23. _____

24. 1.7

24. _____

25. $\dfrac{2}{3}$

25. _____

26. $\sqrt{14}$

26. _____

Use integers to represent the values in each statement.

27. Since I started my diet, last month, I have lost 6 pounds.

27. _____

28. When balancing her checkbook, Jane found that she was overdrawn by $20.

28. _____

29. The weather man on channel six news promised that there will be an increase of 15 degrees over the next few days.

29. _____

30. During a Chicago Bears game, the quarterback threw a pass, to help the team gain 45 yards.

30. _____

Objective 4
Find the absolute values of each number.

31. $|-5|$ 31. _____

32. $|78|$ 32. _____

33. $|-8.9|$ 33. _____

34. $\left|-\dfrac{6}{11}\right|$ 34. _____

Insert $<$, $>$, or $=$ in the appropriate space to make a true statement.

35. $|-3|$ ___ 3 35. _____

36. $|6|$ ___ $|-7|$ 36. _____

37. $|0|$ ___ $|-8|$ 37. _____

38. $\dfrac{45}{9}$ ___ $\dfrac{-60}{6}$ 38. _____

Tell whether the statement is true or false. If false, state why.

39. Every whole number is a real number.   39. _____

40. Every real number is a rational number.   40. _____

41. Every integer is a whole number.   41. _____

Concept Extension

Use the following table to answer the remaining questions.

According to MSN Weather, the chart below tells us the record low temperatures for the Windy City of Chicago, IL.

Record Low Temperatures for Chicago, IL			
Month	Temperature (°F)	Month	Temperature (°F)
January	-24°	July	51°
February	-12°	August	49°
March	0°	September	37°
April	32°	October	24°
May	32°	November	0°
June	41°	December	-20°

42. Which month has the lowest temperature on record?   42. _____

43. Which month has the highest temperature on record?   43. _____

44. Put the months in order from lowest to highest temperature. 44. _____

Section 1.3   Fractions

> **Learning Objectives**
> 1. Write fractions in simplest form.
> 2. Multiply and divide fractions.
> 3. Add and subtract fractions.

Vocabulary.
Use these words to complete the following sentences.

Composite	Factors	Product
Denominator	Numerator	Reciprocals
Equivalent	Prime	Simplified

1. Fractions that are on the same spot on the number line are called _____.

2. In a fraction, the top is called the _____ and the bottom is called the _____.

3. When a fraction is reduced down to it lowest terms, it is said to be _____.

4. Every _____ number can be written as a product of _____ numbers.

5. If the product of two fractions equals one, the fractions are said to be _____.

6. The result of multiplying a list of _____ is called a _____.

Objective 1.3.1
Write each number as a product of primes.

7. 16   7. _____

8. 40   8. _____

9. 120   9. _____

10. 625   10. _____

Write the fraction in lowest terms.

11. $\dfrac{35}{70}$ 11. _____

12. $\dfrac{18}{32}$ 12. _____

13. $\dfrac{24}{56}$ 13. _____

14. $\dfrac{21}{91}$ 14. _____

Objective 2
Multiply or divide as indicated. Write the answer in lowest terms.

15. $\dfrac{4}{3} \cdot \dfrac{6}{7}$ 15. _____

 16. $\dfrac{2}{3} \cdot \dfrac{3}{4}$ 16. _____

17. $\dfrac{8}{9} \div \dfrac{24}{27}$ 17. _____

18. $2\dfrac{1}{4} \div 6\dfrac{1}{3}$ 18. _____

Objective 3
Add or subtract as indicate. Write the answer in lowest terms.

19. $\dfrac{7}{8} - \dfrac{5}{8}$ 19. _____

 20. $\dfrac{17}{21} - \dfrac{10}{21}$ 20. _____

21. $\dfrac{15}{144} + \dfrac{27}{144}$ 21. _____

Write each fraction as an equivalent fraction with the given denominator.

22. $\dfrac{5}{9}$ with a denominator of 45.   22. _____

23. $\dfrac{3}{10}$ with a denominator of $30x$.   23. _____

Add or subtract as indicated. Write the answer in lowest terms.

24. $\dfrac{4}{7} + \dfrac{5}{9}$ 24. _____

25. $4 - \dfrac{10}{11}$ 25. _____

26. $1\dfrac{2}{3} + 3\dfrac{2}{3}$ 26. _____

Concept Extension
Solve.

27.   Susie has $\dfrac{2}{3}$ teaspoons of nutmeg. Last night for dessert she used $\dfrac{1}{4}$ teaspoon. Tonight she

wants to make a pumpkin pie. The recipe calls for $\dfrac{1}{2}$ teaspoon. Does Susie have enough

nutmeg?

                                                        27. _____

Section 1.4   Introduction to Variable Expressions and Equations

Learning Objectives

    1. Define and use exponents and the order of operations.
    2. Evaluate algebraic expressions, given replacement values for variables.
    3. Determine whether a number is a solution of a given equation.
    4 Translate phrases into expressions and sentences into equations.

Vocabulary.

Use these words to complete the following sentences.

Add	**Expression**	**Solution**
Base	**Grouping**	**Subtract**
Exponent	**Multiply**	**Variable**

1.   In the expression $2\left[3+(5-4)\right]$ you will _____ first.

2.   In the expression $2\left[3+(5-4)\right]$ you will _____ second.

3.   In the expression $2\left[3+(5-4)\right]$ you will _____ last.

4.   A(n) _____ is a mathematical statement that two expressions are equal.

5.   A numerical value that makes an equation true is the _____ .

6.   Examples of _____ symbols are ( ), [ ] , and { }.

7.   In the exponential expression: 4^5, 4 is the _____ and 5 is the _____.

8.   A(n) _____ is a symbol or letter that represents a number.

Objective 1

Evaluate.

9. 3^3 9. _____

10. $\left(\dfrac{4}{5}\right)^2$ 10. _____

11. 0.02^5 11. _____

Simplify each expression.

12. $3\bullet 4-5\bullet 2$ 12. _____

13. $6^2+12\bullet 3-8\div 4$ 13. _____

                                                 Martin-Gay *Beginning Algebra, Fifth Edition*

14. $2\left[5+2(8-3)\right]$ 14. _____

15. $\dfrac{18+(12-4)^2}{10^2}$ 15. _____

Evaluate each expression when x = 2, y = 3, and z = 5.

16. $4xyz$ 16. _____

17. $\dfrac{z}{3x}$ 17. _____

18. $\left| y^2 - x^3 \right|$ 18. _____

19. $\dfrac{z^3 + x^2}{y(3+z)}$ 19. _____

Objective 3
Determine whether the given number is a solution of the given equation.

20. Is 4 a solution of $3x+8=5x$?   20. _____

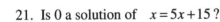 21. Is 0 a solution of $x=5x+15$?   21. _____

22. Is 2 a solution of $2x=4x-5$?   22. _____

Objective 4
Write each phrase as an algebraic expression. Let x represent the unknown number.

23. Two-thirds times a number increased by six.   23. _____

24. The difference of twice a number and nine.   24. _____

25. The product of three and the sum of a number and eight.   25. _____

Write each sentence as an equation or inequality. Use *x* to represent any unknown number.

26. The quotient of eight and a number is not equal to two.   26. _____

27. Five less a number is greater than or equal to negative seven.   27. _____

28. Twice the difference of two and a number yields nine.   28. _____

Concept Extensions

29. Explain why the square of the sum of two numbers is different than the sum of the squares of two numbers.

Section 1.5 Adding Real Numbers

Learning Objectives
1. Add real numbers with the same sign.
2. Add real numbers with unlike signs.
3. Solve problems that involve addition of real numbers.
4. Find the opposite of a number.

Vocabulary.
Use these words to complete the following sentences.

0	***-n***	**Opposites**
n	**Negative number**	**Positive number**

1. _____ are numbers that are equidistant from zero on the number line, but are on

    opposite sides.

2. The result you get from adding two opposites together is _____.

3. If *n* is a negative number, then − (-*n*) is _____.

4. The result of adding -5 and 3 will result in a(n) _____ .

Objective 1
Add.

5. $64 + 739 + 165$ 5._____

6. $-25 + (-56)$ 6._____

7. $-4.7 + (-7.8)$ 7._____

Objective 2
Add.
8. $86 + (-145)$ 8._____

9. $|-6| + (-61)$ 9._____

10. $-25 + 63 + (-45)$ 10._____

11. $6.7 + (-3.7) + 8.9 + (-9.2)$ 11._____

12. $[-56+41]+[78+(-102)]$ 12. _____

Objective 3
Solve.

13. The overnight low temperature in New York was $-15°F$ for Sunday night. The temperature went up $27°F$ by lunch time on Monday. What was the temperature on Monday afternoon?

13. _____

14. A frog fell down a well that was 15 feet deep. He was able to jump and land on a brick that was 3 feet from the bottom of the well. How far from the top of the well is the frog.

14. _____

Objective 4
Find each additive inverse or opposite.

15. 8   15. _____

16. -2 16. _____

17. $|-17|$ 17. _____

Simplify each of the following.

18. $-(-15.67)$ 18. _____

19. $-|5|$ 19. _____

20. $-\left|-\dfrac{4}{7}\right|$ 20. _____

Concept Extension

21. The expression $-x + y$ can be read as "the opposite of x plus y." Write each of the following expression into words.

    a. $x + (-y)$ 21 a. _____

    b. $-x + (-y)$ 21 b. _____

    c. $-x - y$ 21 c. _____

Section 1.6 Subtracting Real Numbers

Learning Objectives
1. Subtract real numbers.
2. Add and subtract real numbers.
3. Evaluate algebraic expressions using real numbers.
4. Solve problems that involve subtraction of real numbers.

Vocabulary.
Use the choices to complete each statement.

23 – x	**Add**	**Opposite**
x – 23	**Complementary**	**Supplementary**

1. Angles whose sum is equal to 180° are called _____ angles.

2. _____ is the expression for 23 less a number.

3. Angles whose sum is equal to 90° are called _____ angles.

4. A number subtracted from 23 is the expression: _____.

5. A number decreased by 23 is the expression: _____.

6. To find the difference of two numbers you can _____ the first number to the

    _____ of the second number.

7. _____ is the expression for a number minus 23.

Objective 1
Subtract.

8. $-5-9$ 8. _____

 9. $16-(-3)$ 9. _____

10. $6.7-(-12.4)$ 10. _____

11. $-\dfrac{3}{5}-\dfrac{4}{7}$ 11. _____

Perform the operation.

12. Decrease 9 by -14.

12. _____

13. Subtract 56 from 29.

13. _____

14. 15 less -31.

14. _____

Objective 2

Simplify each expression.

 15. $-10-(-8)+(-4)-20$

15. _____

16. $16-24+(-32)-19$

16. _____

17. $(-2)^3 + \left[2(6-10)\right]$

17. _____

18. $|12-19|+4^2 -\left[8-11\right]^2$

18. _____

Objective 3

Evaluate each expression when $x = 2$, $y = -4$, and $z = -7$.

19. $x - z$

19. _____

20. $\dfrac{12+y}{z-5}$

20. _____

21. $z^2 - y^3$

21. _____

Objective 4
Solve.

22. A commercial jet liner hits an air pocket and drops 250 feet. After climbing 120 feet, it drops another 178 feet. What is its overall vertical change?

22. _____

23. Joe shot rounds of 3 over par, 2 under par, 3 under par, and 5 under par for a four day tournament of golf. What was his final score at the end of the fourth day?

23. _____

24. Sarah opened a checking account with a deposit of $101.37. She then wrote checks and had other credits against her account. What is her current balance?

Deposit		101.37
Wal-mart	65.29	
Target	54.89	
ATM	60.00	
ATM Card Fee	3.50	

24. _____

Find each unknown complementary or supplementary angle.

25.

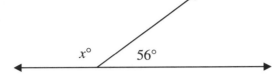

25. _____

26.

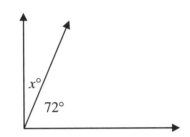

26. _____

Concept Extensions

27. If *m* is positive number and *n* is negative number, determine whether each statement is true or false. If it is false, show an example showing that it is false.

    a. $m - n$ is positive.   27 a. _____

    b. $n - m$ is positive.   27 b. _____

    c. $|m| - |n|$ is always negative.   27 c. _____

    d. $|n| - |m|$ is always a positive number.   27 d. _____

Section 1.7 Multiplying and Dividing Real Numbers

Learning Objectives
1. Multiply and divide real numbers
2. Evaluate algebraic expressions using real numbers.

Vocabulary.
Use the choices to complete each statement.

0	Negative	Reciprocal
1	Positive	Undefined

1. The result of multiplying two negative numbers will be _____.

2. When zero is the divisor, the result will be _____.

3. Another word for multiplicative inverse is _____.

4. When multiplying real numbers, if you have an odd number of negative factors, the final answer will

    be _____.

5. The quotient of a positive and a negative number will be a _____ number.

6. When zero is the dividend, the result will be _____.

7. The number _____ does not have a multiplicative inverse.

Objective 1
Multiply.

8. $-8(9)(-2)$ 8. _____

 9. $\dfrac{2}{3}\left(-\dfrac{4}{9}\right)$ 9. _____

10. $(-0.2)(-4.3)(-2.2)$ 10. _____

11. $4(-2)(-3)(5)(-1)$ 11. _____

Perform the indicated operation.

12. $(-3)(5)-4(-4)$ 12. _____

13. $8(-3)(2)+(-2)(-7)-(-12)(3)$ 13. _____

Evaluate.

14. $(-4)^3$ 14. _____

15. -2^5 15. _____

16. $(-8)^2$ 16. _____

17. -8^2 17. _____

Find the reciprocal or multiplicative inverse.

 18. -14 18. _____

19. $\dfrac{5}{-6}$ 19. _____

20. $\dfrac{1}{3.4}$ 20. _____

Divide.

21. $24 \div (-4)$ 21. _____

22. $\dfrac{0}{-12}$ 22. _____

23. $-\dfrac{5}{9} \div \left(-\dfrac{3}{4}\right)$ 23. _____

24. $64 \div 0$ 24. _____

Objective 2

If $x = -4$ and $y = 3$, evaluate each expression.

25. $5x - 8y$ 25. _____

26. $\dfrac{3y + 12x}{-13}$ 26. _____

27. $2y^3 + y^2 - y - 6$ 27. _____

Concept Extensions.

Simplify.

28. $\dfrac{\left|-8+6\right|^3 + \left|-3-7\right|^2}{\left[2\left(4-3^2\right)+4\right]}$

28. _____

29. List any real numbers that are their own reciprocal.

29. _____

Section 1.8 Properties of Real Numbers

Learning Objectives
1. Use the commutative and associative properties.
2. Use the distributive property.
3. Use the identity and inverse properties.

Vocabulary.
Use the choices to complete each statement.

Additive inverses   **Associative property of multiplication**

Associative property of addition.   **Commutative property of multiplication**

Commutative property of addition   **Distributive property**

Identity element of addition   **Identity element of multiplication**

1. $6 + z = z + 6$ is an example of the _____.

2. $5(xy) = (5x)y$ is an example of the_____.

3. If the sum of two numbers is 0, they are said to be _____.

4. $9 + 0 = 9$ is an example of the _____

5. $(2 + x) + w = 2 + (x + w)$ is an example of the _____.

6. $4 \bullet h = h \bullet 4$ is an example of the _____.

7. $6 \bullet 1 = 6$ is an example of the _____.

8. $3(x - 7) = 3x + 21$ is an example of the _____.

Objective 1
Use commutative property to complete each statement.

 9. $x + 16 =$_____   9. _____

10. $6 \bullet p =$_____   10. _____

11. $5j + 3k =$ _____   11. _____

Use associative property to complete each statement.

12. $5 + (d + 4) =$ _____   12. _____

13. $(10c)s =$ _____   13. _____

14. $(-6 + r) + t =$ _____   14. _____

Use the commutative and associative properties to simplify each expression.

15. $(4+b)+9$ 15. _____

16. $8(7g)$ 16. _____

17. $-\dfrac{8}{13}\left(s\bullet\dfrac{39}{16}\right)$ 17. _____

Objective 2

Use the distributive property to write each expression without parentheses.

18. $3(x-9)$ 18. _____

19. $-(r-3-7p)$ 19. _____

20. $\dfrac{3}{4}(8x-16y)$ 20. _____

21. $5x+7(10-4x)$ 21. _____

Use the distributive property to write each sum as a product.

22. $8\bullet x+8\bullet 2$ 22. _____

23. $11x+11y$ 23. _____

24. $-5t-25$ 24. _____

25. $\dfrac{1}{4}t+\dfrac{1}{2}s$ 25. _____

Objective 3

26. Which of the following $\left\{1, -1, 0, \dfrac{5}{6}, -\dfrac{5}{6}, \dfrac{6}{5}, -\dfrac{6}{5}\right\}$ is the

    a. Reciprocal of $-\dfrac{5}{6}$

                                                 26 a. _____

    b. Opposite of $\dfrac{6}{5}$

                                                 26 b. _____

Concept Extensions

27. Give an example to show that division does not follow the commutative property.

                                                 27. _____

28. Simplify the expression $16 - (12 - 9)$ and the expression $(16 - 12) - 9$ to show that subtraction does not follow the rules of the associative property.

                                                 28. _____

Chapter 1 Vocabulary

Vocabulary Word	Definition	Example
Set	Collection of elements, enclosed in braces.	$\{1, 2, 3, \ldots\}$
Base	The repeated factor of an exponential expression	2^5; 2 is the base.
Denominator	The bottom number of a fraction.	$\dfrac{2}{5}$; 5 is the denominator.
Equation	A mathematical statement that two expressions are equal.	$3x + 4 = 16$
Exponent	The number of times that the base is repeated in an exponential expression.	2^5; 5 is the exponent
Grouping symbols		$\{\ \}, [\ ], (\ )$
Inequality symbols		$<, >, \leq, \geq, \neq$
Numerator	The top number of a fraction.	$\dfrac{2}{5}$; 2 is the numerator
Opposites	Two numbers that are the same distance from 0, but are on opposite sides of 0 on the number line.	18 and -18 are opposites.
Reciprocals	Two numbers whose product is 1.	$\dfrac{3}{4}$ and $\dfrac{3}{4}$ are reciprocals.
Solution	A value of a variable that makes the equation true.	$x + 7 = 19$ 12 is a solution. $12 + 7 = 19$
Variable	A symbol or letter that represents a number.	x, y, or z

Chapter 1 Practice Test A

Translate the statement into symbols.

1. Eight more then twice a number is less than negative fifteen.

2. Subtract half a number from twenty-one.

Simplify the expression.

3. $(-18)(-9)$

4. $\dfrac{5}{11} - \left(-\dfrac{3}{7}\right)$

5. $14 \div 2 \bullet (-4) + 16 \div 4$

6. $-3\dfrac{2}{3} \bullet \left(-5\dfrac{1}{4}\right)$

7. $\dfrac{0}{12}$

8. $\dfrac{|12-22|}{|18-9|}$

9. $-17 + (-13) + 19$

1. _____

2. _____

3. _____

4. _____

5. _____

6. _____

7. _____

8. _____

9. _____

10. $(7-10)^3 - 9(2)^2 + 20(-1)$

10. _____

11. $|12-19| + 3[9-2(5+2)]$

11. _____

Insert $<, >$, or $=$ in the appropriate space to make the statement true.

12. -18 _____ 12

12. _____

13. $|-22|$ ____ $|3(8)|$

13. _____

14. -2.756 _____ -2.75600

14. _____

15. Given the set $\left\{-12, -4, 0, \dfrac{5}{9}, 1, \pi, \sqrt{25}, 11.75\right\}$, list the numbers that also belong to the set of:

    a. Natural Numbers

15 a. _____

    b. Whole Numbers

15 b. _____

    c. Integers

15 c. _____

    d. Rational Numbers

15 d. _____

    e. Irrational Numbers

15 e. _____

    f. Real numbers

15 f. _____

16. Evaluate $2x^2 - 3x - 1$ if $x = -3$.

16. _____

Identify the property illustrated by each expression.

17. $0 + t = t$

17. _____

18. $(5+n) + m = 5 + (n+m)$

18. _____

19. $\dfrac{4}{9} \cdot 1 = \dfrac{4}{9}$

19. _____

20. $-6(7-m) = -42 + 6m$

20. _____

21. Find the multiplicative inverse of -15.

21. _____

22. Find the opposite of $-\dfrac{5}{7}$.

22. _____

Solve.

23. Over a seven day period the Dow Jones Industrial Average has gains of 32, 24, and 18 points. In that period it had loses of 29, 37, 15, and 21 points. Find the average daily performance over then seven day period.

23. _____

24. Two contractors bid on a home remodeling a house. The first bids $9,350 for the entire job. The second contractor will work for $25.50 per hour, plus $4,200 for materials. He estimates that the job will take 220 hours. Which contractor has the lower bid?

24. _____

25. A student saved $18,000 to attend graduate school. If she estimates that her expenses will be $723.50 a month while in school, does she have enough to complete her 18-month master's degree program?

25. _____

Chapter 1 Practice Test B

Translate the statement into symbols.

1. The quotient of a number and negative three is not equal to five.

   a. $\dfrac{-3}{x} \neq 5$ b. $-3x \neq 5$ 1. _____

   c. $\dfrac{x}{-3} \neq 5$ d. $x(-3) \neq 5$

2. The difference of sixteen and a number is less than twelve.

   a. $16 - x < 12$ b. $x - 16 < 12$ 2. _____

   c. $16 - x > 12$ d. $x - 16 > 12$

Simplify the expression.

3. $(-1440) \div 12$

   a. -240   b. -120   3. _____

   c. -12   d. -144

4. $\left(-\dfrac{6}{7}\right) - \left(-\dfrac{2}{9}\right)$

   a. $-\dfrac{68}{63}$ b. $-\dfrac{1}{4}$ 4. _____

   c. $-\dfrac{1}{2}$ d. $-\dfrac{40}{63}$

5. $\left[-12 \div 3(7 - 10)\right] + (-2)^3$

   a. -4   b. 4   5. _____

   c. $-\dfrac{20}{3}$ d. -20

6. $-6.724 \bullet (-3.25)$

   a. 21.853   b. -9.974   6. _____

   c. -3.474   d. 2.0689

7. $\dfrac{-5}{0}$

     a. 0   b. $\varnothing$ 7. _____

     c. -5   d. 5

8. $\left|19-4^2+(-14)\right|-\left|13-22\right|$

     a. -2   b. 2   8. _____

     c. -31   d. -9

9. $-(7)^3-(8-18)^3$

     a. -657   b. -1343 9. _____

     d. 657   d. 1343

10. $2+3\bullet\left|4-(7^2-6^2)\right|$

     a. 45   b. -25 10. _____

     c. -45 d. 29

11. $\dfrac{3^2-2\bullet4}{-30+2\bullet4^2}$

     a. $-\dfrac{1}{16}$ b. $\dfrac{1}{2}$ 11. _____

     c. 14   d. $\dfrac{1}{4}$

Insert <, > , or = in the appropriate space to make the statement true.

12. -2^3 ____ $(-2)^3$

     a. >   b. <   c. =   12. _____

13. $\left|6-4^2\right|$ ____ $\left|-20+2(5-2)\right|$

     a. >   b. <   c. =   13. _____

14. $-\dfrac{1}{4}$ ____ -0.245

     a. >   b. <   c. =   14. _____

15. Tell which set or sets the number -5 belongs to.

     a. Integer, Rational, Real   b. Whole, Real   15. _____

     c. Irrational, Real   d. Real

16. Evaluate $b^2 - 4ac$ if $a = -2$, $b = -5$ and $c = 7$

     a. 31   b. 81   16. _____

     c. -81 d. -31

Identify the property illustrated by each expression.

17. $-3 \bullet 1 = -3$ 17. _____

     a. Identity element of addition   b. Inverse element of addition

     c. Identity element of multiplication   d. Inverse element of multiplication

18. $(5p)q = 5(pq)$ 18. _____

     a. Associative property of multiplication   b. Distributive property

     c. Commutative property of multiplication   d. Associative property of addition

19. $-(3z - 5y) = -3z + 5y$ 19. _____

     a. Associative property of multiplication   b. Commutative property of multiplication

     c. Distributive property   d. Associative property of addition

20. $1 + 0 = 1$

        20. _____

    a. Identity element of addition   b. Inverse element of addition

    c. Identity element of multiplication   d. Inverse element of multiplication

21. Find the reciprocal of $\dfrac{-13}{14}$.

    a. $\dfrac{14}{13}$ b. $\dfrac{-14}{13}$ 21. _____

    c. $\dfrac{13}{14}$ d. $-\dfrac{13}{14}$

22. Find the additive inverse of $-(-4)$

    a. 4   b. -4 22. _____

    c. $\dfrac{1}{4}$ d. $-\dfrac{1}{4}$

Solve.

23.   Four students recorded time they spent working on a take-home exam: 5.2, 4.7, 9.5, and 8 hours. Find the average time spent.

    a. 6.75 hours   b. 6.85 hours

    c. 7 hours   d. 6.50 hours

                                                   23 _____

24. How many pieces of pipe that are $\dfrac{2}{3}$ foot long must be laid together to make a pipe that is 16 feet long?

    a. 11 pieces   b. 24 pieces

    c. 32 pieces   d. 48 pieces

                                                   24. _____

25. The expression $2L + 2W$ gives the perimeter of a rectangle with the length, L and the width, W. Find the perimeter of a rectangle whose length is $16\frac{2}{3}$ feet and width is $7\frac{3}{4}$ feet.

a. $32\frac{5}{12}$ feet

b. $48\frac{5}{6}$ feet

c. $24\frac{5}{12}$ feet

d. $36\frac{5}{6}$ feet

25. _____

Chapter 2 Equations, Inequalities, and Problem Solving
Section 2.1 Simplifying Algebraic Expressions

Learning Objectives
 1. Identify terms, like terms, and unlike terms.
 2. Combine like terms.
 3. Use the distributive property to remove parentheses.
 4. Write word problems as algebraic expressions.

Vocabulary.
Use the choices to complete each statement.

Combine like terms   **Numerical coefficient**   **Like terms**

Terms   **Unlike terms**   **Exponent**

1. A product of a number and variables raised to powers is called a(n) _____.

2. Terms that have the same variables raised to the same power are called _____.

3. In the term $3x^2$, the three is called the _____.

4. To simplify the expression $8t + 10t$ we _____.

Objective 1

5. Complete the following table for the expression $3x^4 + \dfrac{1}{5}x^3 - x + 7$

Term	Numerical Coefficient	Variable	Exponent

6. Determine whether the following are like or unlike terms

   a. $5x$, $\dfrac{4}{25}x$ 6a. _____

   b. $17xy^2$, $27x^2y$ 6b. _____

Objective 2
Simplify each expression by combining like terms.

7. $17t + 3s - 29t - 9s$ 7. _____

8. $8x^3 + x^3 - 11x^3$ 8. _____

9. $1.5x - 2.5xy + 8.8yx - 9x$ 9. _____

Objective 3

Simplify each expression. First use the distributive property to remove any parentheses.

10. $6(t-5)$ 10. _____

11. $-\dfrac{1}{4}(12y - 8x + 16z)$ 11. _____

 12. $5(x+2) - (3x-4)$ 12. _____

13. $0.6(t-3.2) - (1.4 - 3.7y)$ 13. _____

Objective 4

Write each of the following as an algebraic expression. Simplify if possible.

14. Add $12x - 8$ and $-16z + 19$.   14. _____

15. Subtract $3z - 9y$ from $18y - 5z$.   15. _____

Write each of the following as an algebraic expression. Simplify if possible. Let x represent the unknown number.

16. Eight times a number increased by half a number less twelve.

16. _____

17. The sum of twice a number, eighteen, and negative five times a number.

17. _____

18. The difference of a number and four, subtracted from the sum of nine and twice a number.

18. _____

Concept Extension

19.   If the value of z was doubled, what would happen to the value of 17z?

19. _____

20. If both c and d were doubled, what would happen to the value of $7cd^2$?

20. _____

Section 2.2 The Addition Property of Equality

Learning Objectives
1. Define linear equations and use the addition property of equality to solve linear equations.
2. Write word phrases to algebraic expressions.

Vocabulary.
Use the choices to complete each statement.

| Addition | Equation | Expression |
| Solution | Solving | True |

False

1. The process to find a solution is referred to as _____ an equation for the variable.

2. True or false: $x = 3$ is equivalent to $3 = x$. _____

3. A(n) _____ contains an equal sign, while a(n) _____ does not.

4. A(n) _____ is a value for the variable that makes the equation true.

5. $x + 2 = 9$ and $x + 2 - 2 = 9 - 2$ are equivalent equations by the _____ property of

   equality.

Objective 1

Solve each equation.

6. $5 + x = 19$ 6. _____

7. $z - 5.6 = -19.5$ 7. _____

8. $3x = 25 - 2x$ 8. _____

 9. $5b - 0.7 = 6b$ 9. _____

10. $7y - 18 + (-6y) - 13 = 27$ 10. _____

11. $-\dfrac{4}{7}x - \dfrac{3}{4} = \dfrac{3}{7}$ 11. _____

12. $3(x-2)-2(x+5)=19$ 12. _____

 13. $13x-9+2x-5=12x-1+2x$ 13. _____

14. $12.6-2.5x-6.7=-1.5(x-6)$ 14. _____

Objective 2

Solve.

 15. Two numbers have a sum of 20. If one number is p, express the other number in terms of p.

15. _____

16. A 40 foot board was cut into two pieces. If one of the pieces measured to be x feet, express the length of the second piece in terms of x ?

16. _____

17. Two angles are complementary if the sum of the two angles is 90 degrees. If one angle is w degrees, express the measure of its complementary angle in terms of w.

17. _____

18. A rope is cut into 8 equal pieces. If the length of the rope was t feet initially, how long is each piece in terms of t ?

18. _____

19. Mary and Joe combined their money to buy their mom a $200 DVD player. If Mary had m dollars, express the amount of money Joe had in terms of m.

19. _____

Martin-Gay *Beginning Algebra, Fifth Edition*

Concept Extensions

20. Distinguish the difference between 5 less a number and 5 is less than a number.

20. _____

21. A football costs $x and flags cost $y. How much would it cost to buy 6 footballs and 16 flags to prepare for flag football season?

21. _____

Section 2.3 The Multiplication Property of Equality

Learning Objectives
1. Use the multiplication property of equality to solve linear equations.
2. Use both the addition and multiplication properties of equality to solve linear equations.
3. Write word phrases as algebraic expressions.

Vocabulary.

Use the choices to complete each statement.

Addition   **Multiplication**

True   **False**

1. By the _____ property of equality, $3x = 9$ and $\dfrac{3x}{3} = \dfrac{9}{3}$ are equivalent.

2. True or false. To solve for x in the equation $\dfrac{9x}{7} = 3$, the next step would look like this: $\dfrac{9x}{7} \bullet \dfrac{7}{9} = 3 \bullet \dfrac{7}{9}$.

   _____

3. By the _____ property of equality, $\dfrac{5}{9}x = 6$ and $5x = 54$ are equivalent.

Objective 1

Solve each equation.

4. $-5x = 20$ 4. _____

5. $-t = -56$ 5. _____

6. $\dfrac{-7}{9}r = \dfrac{14}{27}$ 6. _____

7. $\dfrac{v}{6} = 0$ 7. _____

8. $4.5x = -16.875$ 8. _____

Objective 2

Solve.

9. $5x - 24 = 101$

9. _____

10. $-x - 19 = -34$

10. _____

 11. $8x + 20 = 6x + 18$

11. _____

12. $\dfrac{1}{5}(15x - 20) = \dfrac{-4}{7} + \dfrac{7}{5}$

12. _____

13. $-x - 3 - 8x = 2(5 - x)$

13. _____

Objective 3

Write each algebraic expression described. Simplify if possible.

14. If x is the first of three consecutive odd integers, express the sum of the first and third integer in terms of x.

14. _____

15. If x is the first of four consecutive integers, express the sum of the third integer and 25 in terms of x.

15. _____

16. If the three sides of a triangle are consecutive even integers with s being the smallest side, express the sum of the perimeter of that triangle in terms of s.

16. _____

17. If the width of a rectangle is x and the length is the next consecutive even integer from x. What is the perimeter of the rectangle in terms of x.

17. _____

18. The numbers on the right side of the hallway, of an apartment building are consecutive even integers. If there are eight doors along that side of the hall, express the sum of the second door and the last door in terms of x if the first door number is x.

18. _____

Concept Extensions

19. If the equation $5x - 16 + 3x = \dfrac{16x - 32}{2}$ is true for all real numbers. Plug in three values for x and describe what happens.

19. _____

20. The equation $(x + 4)^2 = x^2 + 8x - 16$ has no solution. Plug in three values for x and describe what happens.

20. _____

Section 2.4 Solving Linear Equations

Learning Objectives
1. Apply a general strategy for solving a linear equation
2. Solve equations containing fractions.
3. Solve equations containing decimals.
4. Recognize identities and equations with no solutions

Objective 1

Solve each equation.

1. $-5y + 12 = 3(4y - 8)$ 1. _____

2. $5(2x - 1) - 2(3x) = 1$ 2. _____

3. $-(14x - 18) + 2(8x + 16) = 4(7)$ 3. _____

4. $5 - (2x + 6) = -5 + 6(4x - 3)$ 4. _____

5. $8(p - 2) = 5(p + 4) - p$ 5. _____

Objective 2

Solve each equation.

6. $\dfrac{x}{3} + 2 = -9$ 6. _____

7. $\dfrac{2x}{5} + \dfrac{3}{5} = \dfrac{9}{10}$ 7. _____

8. $\dfrac{x}{2} - 1 = \dfrac{x}{5} + 2$ 8. _____

9. $\dfrac{5}{8}x - \dfrac{3}{4} = \dfrac{11}{12}x$ 9. _____

10. $\dfrac{5x - 12}{6} = x + 9$ 10. _____

Objective 3

Solve each equation.

11. $0.5x - 1.2 = 0.35x$ 11. _____

12. $0.50x + 0.15(70) = 35.5$ 12. _____

13. $-0.04x + 0.25(x - 4) = 0.5(x - 2)$ 13. _____

14. $0.04(4y - 2) = 0.02(2 + 3y) + 8$ 14. _____

Objective 4

Solve each equation.

  15. $2(x+3)-5=5x-3(1+x)$

15. _____

16. $3(x-3)+5x+16=4(3x-5)-2(2x+1)$

16. _____

17. $0.02x+0.03(x+4)=0.12(x+1)-0.7x$

17. _____

18. $\dfrac{1}{5}x-\dfrac{1}{2}=\dfrac{1}{20}(4x-5)-\dfrac{3}{4}$

18. _____

Concept Extension

19. Solve the equation: $0.80(x+5)=\dfrac{3}{2}(x-2)$.

19. _____

20. The perimeter of a quadrilateral is 124 inches. The sides of the quadrilateral are $x,\ 2.5x,\ \dfrac{3}{4}(x-2),$ and $10+x$. Find the length of all four sides.

20. _____

Section 2.5 An Introduction to Problem Solving

Learning Objectives
1. Solve problems involving direct translations.
2. Solve problems involving relationships among unknown quantities.
3. Solve problems involving consecutive integers.

Objective 1

Write each of the following as equations. Then solve.

1. Twice a number increased by eight amounts to sixteen. Find the number.

1. _____

2. Fourteen times a number increased by seven yields seven times the sum of one and twice a number. Find the number.

2. _____

3. Twice the difference of a number and 8 is equal to three times the sum of the number and 3. Find the number.

3. _____

4. If the sum of a number and seven is tripled, the result is eight times a number.

4. _____

5. The sum of two and three times the difference of a number and five is equal to four times the difference a number and one.

5. _____

Objective 2

Solve.

6. A 24 foot board is cut into two pieces so that the smaller piece measures x feet. If the longer piece is 6 feet less than twice the shorter piece, what is the length of both sides?

6. _____

7. In 1986, the Chicago Bears beat the New England Patriots by 36 points in the Super Bowl. If a total of 56 points were scored, what was the final score of the game? (Source: National Football League)

7. _____

Martin-Gay *Beginning Algebra, Fifth Edition*

8. A sales clerk's sales in August were $2500 less than 2 times her sales of June. If his August sales were $9000, by what amount did her sales increase?

8. _____

9. The perimeter of a square piece of land is twice the perimeter of an equilateral triangle. (An equilateral triangle has three equal sides.) If one side of the square is 45 feet, find the length of the sides of the triangle.

9. _____

10. Two angles are supplemental if their sum is 180° The larger angle measures eight degrees more than three times the measure of the smaller angle. If x represents the measure of the smaller angle and these two angles are supplementary, find the measure of each angle.

10. _____

11. The width of a rectangular swimming pool is 13 meters less than the length and the perimeter is 98. Find it dimensions.

11. _____

12. According to People Magazine, Angelina Jolie and Halle Berry are in the top ten lists of highest paid actresses. Halle average salary is one-third the amount of Angelina. If combined they make $30,000, what is the average salary of each actress.

12. _____

13. A plumber cuts a 75 inch pipe into 3 pieces. If the longest piece is three times the smallest piece and the middle-sized piece is two more then the smallest. What is the length of all three pieces?

13. _____

Objective 3

Solve.

14. The sum of two consecutive even integers is 138. Find the integers.

14. _____

15. The measures of the angles of a triangle are 3 consecutive even integers. Find the measure of each angle.

15. _____

16. The sum of three consecutive odd integers is -96. Find the integers.

16. _____

17. The ages of three sisters are consecutive even integers. If the sum of three times the youngest, the middle sister's age, and twice the oldest is 58. Find the ages of the three sisters.

17. _____

18. Houses on Main Street have consecutive odd numbers. If the sum of two neighboring houses is 814. What are the numbers of each house?

18. _____

19. There is three consecutive even integers such that three times the second is six more than the sum of the first and third. Find the three integers.

19. _____

Concepts Extensions

20. If two lines intersect as the illustration shows, angle 1 $(\angle 1)$ and $\angle 2$, $\angle 3$, and $\angle 4$, are called vertical angles. Let the measure of $\angle 1$ be any three random angle values. Compute the values of the other three angles. What did you discover?

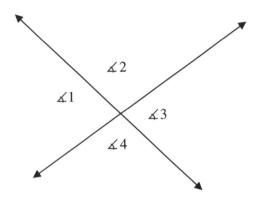

Section 2.6 Formulas and Problem Solving

Learning Objectives
1. Use formulas to solve problems.
2. Solve a formula or equation for one of its variables.

Objective 1

Substitute the given values into each given formula and solve for the unknown variable. If necessary, round to one decimal place.

1. $d = rt$ $d = 256$ $t = 8$ 1. _____

2. $V = l \bullet w \bullet h$ $V = 126$ $w = 3$ $h = 6$ 2. _____

3. $I = P \bullet r \bullet t$; $I = 300$, $P = 1500$, $r = 0.05$ 3. _____

4. $y = mx + b$; $y = -1$, $x = 5$, $b = -16$ 4. _____

5. $F = \dfrac{9}{5}C + 32$; $F = 77$ 5. _____

6. We are planning a trip to Disney World this year. We live 1200 miles away from Kissimmee, FL. The online map website said that it would take us about $18\frac{1}{2}$ hours. How fast does the website expect us to drive?

6. _____

7. Convert San Antonio Texas' highest temperature of 110 degrees Fahrenheit to Celsius. (Source: The Weather Channel)

7. _____

8. An architect designs a rectangular garden such that the width is exactly two-thirds of the length. If 260 feet of antique picket fencing are to be used to enclose the garden, find the dimensions of the garden.

8. _____

9. How long will it take an account that started with $2000 to double, if the account has a simple interest rate of 8.5%? Round your answer up to the next whole year.

9. _____

10. Joe and Janet are driving from their home in separate vehicles to the coast 425 miles away. Jeff leaves at 11:00 AM and averages 55 miles per hour. Janet leaves two hours later at 1:00 PM, but she averages 65 mph. Which person arrives at the coast first?

10. _____

11. If the perimeter of a tennis court is 228 feet and the width is three less than half the length. What are the dimensions of the tennis court?

11. _____

Objective 2

Solve each formula for the specified variable.

12. $A = l \bullet w$ for w.

12. _____

13. $A = \frac{1}{2}bh$ for b.

13. _____

14. $C = 2\pi r$ for r

14. _____

 15. $S = 2\pi rh + 2\pi r^2$ for h.

15. _____

16. $F = \frac{9}{5}C + 32$ for C.

16. _____

Martin-Gay *Beginning Algebra, Fifth Edition*   51

17. $V = \frac{1}{3}\pi r^2 h$ for h.

17. _____

18. $A = \frac{1}{2}(B+b)$ for b.

18. _____

Concept Extensions

19. Suppose that your fellow classmate analyzes a distance problem that follows "Sarah has traveled 45 miles before Mary starts. Because Mary travels 5 mph faster than Sarah, it will take her $\frac{45}{5} = 9$ hours to catch Sarah." How would you react to this analysis of the problem?

19. _____

20. When a car of mass m collides with a wall, the energy of the collision is given by the formula $E = \frac{1}{2}mv^2$. Compare the energy of two collisions: a car striking a wall at 20 mph, and at 40mph.

20. _____

Section 2.7 Percent and Mixture Problem Solving

Learning Objectives
1. Solve percent equations.
2. Solve discount and mark-up problems.
3. Solve percent increase and percent decrease problems.
4. Solve mixture problems.

Objective 1

Find each number described.

1. What number is 120% of 50?

1. _____

2. The number 45 is 25% of what number?

2. _____

3. 160% of what number is 144?

3. _____

4. The number 56.25 is 45% of what number?

4. _____

5. The number 28.6 is what percent of 52?

5. _____

6. 23% of 20 is what number?

6. _____

Objective 2

Solve . If needed round answers to the nearest cent.

7. The dress you have been looking for is on sale for 35% off of the original price. If it originally is $120.00, what is the sale price?

         7. _____

8. The college bookstore will always mark-up the price of textbooks 15% of what the amount they purchased the books from the publishers. If the publisher sells a specific text to the bookstore for $85, what will the bookstore sell that text to the students for?

         8. _____

9. Find the rate of discount if the discount is $50 and the original price was $400.

         9. _____

10. A 18% discount on a big screen TV saved Tony Williams $154.26. What was the original price of the TV?

         10. _____

 11. A birthday celebration meal is $40.50 including tax. Find the total cost if 15% tip is added to the cost.

         11. _____

12. Greg purchased a Jeep Cherokee. The selling price plus the 8.25% state sales tax was $34,640.00. How much tax was added to the price of the vehicle?

         12. _____

Objective 3

Solve. If necessary round to the nearest hundredth.

13. The value of my house has gone up to 110,000 from 99,000. What was the percent increase?

    13. _____

14. The price of gas has gone up 53% in the past three years. If three years ago the price of gas was $1.89 per gallon, what is the current price of gas?

    14. _____

15. Every year we are guaranteed to get a 5% pay raise. If my contract for next year states I make $56,000, what is my salary for this year?

    15. _____

16. It has been stated that the crime rate in your neighborhood is 5% lower this year than it was last year. According to records, last year there were 120 crimes committed in your neighborhood. How many crimes were committed this year?

    16. _____

Objective 4

Solve. If necessary, round your answer to the nearest tenth.

17. Walnuts sell for $4.80 per pound, and pecans sell for $6.40 a pound. How many pounds of pecans should be mixed with 20 pounds of walnuts to get a mixture that can be sold for $5.40?

    17. _____

18. Hawaiian Punch is 10% fruit juice. How much water would you have to add to one gallon of Hawaiian Punch to get a drink that is 6% fruit juice for your infant son.

    18. _____

19. How many liters of a 20% alcohol solution should Jerry mix with 50 liters of a 60% alcohol solution to obtain a 30% solution?

19. _____

20. Gus has on hand a 5% alcohol solution and a 20% alcohol solution. He needs 30 liters of a 10% alcohol solution. How many liters of each solution should he mix together to obtain the mixture he desires?

20. _____

Concept Extensions

21. To calculate a 15 % tip on a $24 bill a customer adds $2.40 and half of $2.40, or $1.20, to get $3.60. Is this method valid? Why or why not?

21. _____

22. When a discount of 15 % is given by a retailer is it the same as giving three successive discounts of 5 % each? Give arguments to support your answers.

22. _____

Section 2.8 Further Problem Solving

Learning Objectives
1. Solve problems involving distance.
2. Solve problems involving money.
3. Solve problems involving interest.

Objective 2.8.1

Solve. Round answers to once decimal point if necessary.

1. Two cars left St. Louis at the same time. One traveled east at 60 mph and the other went west at 50 mph. How long until the two cars are 715 miles apart?

   1. _____

2. Ashley and Brandon are 315 miles apart. Ashley hops in her car and drives 50 mph towards Brandon. At the same time, Brandon hopped in his car and drove toward Ashley. He drove at a rate of 55 mph. How long does it take them to meet?

   2. _____

3. Suppose that Carl was riding a bicycle at 15 mph, he rode 10 miles farther than Michael, who was riding his bike at 14 mph. Carl rode for 30 minutes longer than Michael. How long did Carl ride his bike?

   3. _____

4. How long will it take a bus traveling at 60 miles per hour to overtake the car traveling at 40 mph if the car had a 1.5 hour head start?

   4. _____

5. Two cars start together and head east, one averaging 56 mph and the other averaging 67 mph. How far apart will the cars be in $7\frac{1}{2}$ hours?

   5. _____

6. On a Sunday morning, Erika walked for 2 hours and then ran for 30 minutes. If she ran twice as fast as she walked and she covered 12 miles altogether, then how fast did she run?

6. _____

Objective 2

Solve. Round to the nearest tenth if needed.

7. Bobby has 75 coins in his piggy bank. He had only nickels and dimes. If he counted his money and the total was $4.50, how many of each coin does he have?

7. _____

8. Part of the proceeds from a garage sale was $280 worth of $5 and $10 bills. If there were 20 more $5 bills than $10 bills, find the number of each denomination.

8. _____

9. The manager of an electronics store placed an order for $21,320 worth of DVD players at $160 each and MP3 players at $120 each. If the number of MP3 players that were ordered was three times the number of DVD players, then how many of each did the manager order?

9. _____

10. The cost for a long-distance phone call is $0.45 for the first minute and $0.35 for each additional minute. If the total charge for a long-distance call is $6.40, how many minutes was the call?

10. _____

11. Tabitha is taking guitar lessons. Because she was not dedicated in the beginning of her lessons, Tabitha's parents make her earn the money to pay for her lessons. Her parents pay her $1.25 to do the laundry and $5.00 to mow the yard. In one month, she did the laundry 8 times as much than she mowed the yard. If her parents paid her $45.00 that month, how many times did she mow the yard?

11. _____

12. Danny works part time as a cashier making $7.25 an hour. During one week he worked 32 hours and made $243.84. He realizes he must have gotten a raise that he did not know about. How much per hour was his raise?

12. _____

Objective 3

Solve. If necessary round to the nearest tenth.

13. Frank invests a certain amount of money at 5% interest and $1360 more than that amount at 7%. His total yearly interest was $241.60. How much did she invest at each rate?

13. _____

14. How can $54,000 be invested, part at 8% annual simple interest and the remainder at 10% annual simple interest, so that the interest earned by the two accounts will be equal?

14. _____

15. A sum of $8000 is invested, part of it at 8% and the remainder at 10%. If the interest earned by the 8% investment is $260 less than the interest earned by the 10% investment, find the amount invested at each rate.

15. _____

16. If $4000 is invested at 5% interest, how much should be invested at 8% interest so that the total return for both investments averages 7%?

16. _____

17. How can $8400 be invested, part at 9% and the remaining at 12%, so that the two accounts will produce the same amount of interest?

17. _____

18. Wayne makes $8000 more per year than his wife. Wayne saves 10% of his income for retirement, and his wife saves 8%. If together they save $6416 per year, then how much does each make in one year?

18. _____

Concept Extensions.

Solve.

19. How many liters of pure alcohol must be added to 30 liters of 10% solution to obtain a 30% solution?

19. _____

20. Michelle has $3.90 worth of nickels, dimes and quarters. The number of dimes is 3 less than the number of nickels. The number of quarters is 7 more than the number of dimes. How many of each coin does she have?

20. _____

Section 2.9 Solving Linear Inequalities.

Learning Objectives
1. Define linear inequality in one variable, graph solution sets on a number line, and use interval notation.
2. Solve linear inequalities.
3. Solve compound inequalities.
4. Solve inequality applications.

Vocabulary.
Use the choices to complete each statement.

Linear inequality in one variable   **True**
Compound inequality   **False**

1. True or false. When dividing both sides of an inequality by a negative number, the inequality remains

   the same. _____

2. A _____ contains two inequalities.

3. $ax + b < c$ is the general form of a _____.

Objective 1

Graph each set of numbers given in interval notation. Then write the inequality statement in x describing the numbers graphed.

4. $(3, \infty)$ 4. _____

5. $(-\infty, -1]$ 5. _____

6. $[10, \infty)$ 6. _____

Graph each inequality on a number line. Then write the solutions in interval notation.

 7. $x \le -1$ 7. _____

8. $y > -3$ 8. _____

9. $w \geq 6$

9. _____

Objective 2

Solve each inequality. Graph the solution set and write it in interval notation.

10. $4x < 16$

10. _____

11. $x - 3 \geq -8$

11. _____

 12. $-8x \leq 16$

12. _____

13. $4x - 6 < 3x + 1$

13. _____

14. $8(5 - x) \leq 10(8 - x)$

14. _____

15. $\dfrac{3x - 3}{2} < 2x + 2$

15. _____

Objective 3

Solve each inequality. Graph the solution set and write it in interval notation.

16. $-4 < x \le 7$

16. _____

17. $-12 \le x - 6 < 9$

17. _____

18. $-6 < 3(x - 2) \le 8$

18. _____

19. $-2 < \dfrac{x-4}{6} < 7$

19. _____

Objective 4

Solve.

20. A student has test scores of 70, 89, 76, and 82 points. What must the student get on her fifth test so that her average is at least 80 points?

20. _____

21. Two less three times a number is less than six more than the number.

21. _____

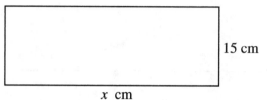 22. Find the values of x so that the perimeter of this rectangle is no greater than 100 centimeters.

15 cm

x cm

22. _____

23. A bride does not want to spend more than $3000 on her wedding reception. If it cost $20 per plate, how many people can she invite to the wedding reception and not go over budget?

23. _____

Concept Extensions.

24. Given the inequality $1 < \dfrac{1}{x}$; is it always true that $x < 1$? Why or why not?

24. _____

Chapter 2 Vocabulary

Vocabulary Word	Definition	Example
Like terms	Terms with the same variable raised to the same power.	5x , 9x, and -4x are like terms.
Numerical Coefficient	The numerical factor of a term	2x 2 is the numerical coefficient
Linear inequality in one variable	An inequality in the form $ax + b < c$ (The < can be replaced by >, ≤ or ≥.	$3x - 6 < 12$
Equivalent equations	Equations that have the same solution.	$x + 2 = 4$ $2x + 6 = 10$
Formula	An equation that describes a known relationship among quantities	$d = rt$
Compound inequality	Inequalities that have two inequality symbols	$-3 < x < 9$
Linear equation in one variable	An equation in the form $ax + b = c$, where a, b, and c are all real numbers.	$3x + 9 = 18$

Chapter 2 Practice Test A

Simplify each of the following expressions.

1. $8t - 19 + (-12t) + 37$

1. _____

2. $3(x - 7) - (2x + 9)$

2. _____

3. $1.4(3.2p - 8) + 4.6 - 7.8$

3. _____

4. $\dfrac{1}{2}x + \dfrac{3}{7}\left(\dfrac{1}{4}x - 6\right)$

4. _____

Solve each of the following equations.

5. $5x + 13 = -3^3$

5. _____

6. $3\left[4x - (5x + 12)\right] = 24$

6. _____

7. $\dfrac{2}{3}x - \dfrac{4}{5} = \dfrac{1}{2}x$

7. _____

8. $0.06x - 0.14 = 0.2(x + 7)$

8. _____

9. $7y - 34 + 9y = 4^2(y - 2)$

9. _____

10. $4(x - 6) - 3(x + 5) = 2(x + 3) + 2(x - 6)$

10. _____

11. $\dfrac{2x - 3}{9} + \dfrac{x + 1}{2} = x - 4$

11. _____

12. $0.08(t - 200) = 44.2 - 0.06t$

12. _____

13. $23 + 4(n - 2) = 9n + 5(3 - n)$

13. _____

Solve each of the following applications.

14. Find the value of B if $b = 19$, and $A = 26$ in the formula . $A = \dfrac{1}{2}(B + b)$

14. _____

15. The quotient of three times a number and 5 amounts to half the number increased by two.

15. _____

16. Find three consecutive odd integers such that the sum of the smallest and four times the largest is 61.

16. _____

17. How much 35% copper alloy should be melted with 37 kg of 75% copper alloy to produce an alloy which is 50% copper. If necessary round your answer to the nearest tenth.

17. _____

18. A bus leaves the station and travels at 50 mph. An hour and a half later, a second bus leaves that same station, but travels at 65 mph. When does the second bus overtake the first?

18. _____

19. The perimeter of a standard football field, including the end zones is 1040 feet. If the field is 360 feet long, how wide is the field?

19. _____

Martin-Gay *Beginning Algebra, Fifth Edition*

Solve each of the following equations for the indicated variable.

20. $S = P + \Pr t$ for t.

20. _____

21. $y = mx + b$ for m.

21. _____

Solve and graph each of the following. Write your answers in interval notation.

22. $2x + 8 > 17$

22. _____

23. $3 - \dfrac{1}{4}x \le 2$

23. _____

24. $2 \le 4 - \dfrac{1}{2}(x - 8) \le 10$

24. _____

Solve.

25. A rectangle's width is three feet more the half its length. If the perimeter of the rectangle is between 12 and 21 feet, what must the measurement of the length be between?

25. _____

Chapter 2 Practice Test B

Simplify each of the following expressions.

1. $7.3r - 2.3 - 3.4r + 6.7 + 2r$
    a. $5.9r - 4.4$ b. $1.9r + 11$

    c. $5.9r + 4.4$ d. $12.9r - 11$

1. _____

2. $4(t-3) - 5(7-t)$
    a. $9t - 23$ b. $-t + 23$

    c. $-t - 47$ d. $9t - 47$

2. _____

3. $-\left[17 + 3(2x-8) - 4(-x)\right]$
    a. $-10x + 7$ b. $-2x + 7$

    c. $10x - 7$ d. $-2x - 41$

3. _____

4. $\dfrac{5}{6}(x-2) - \dfrac{7}{8}(x+4)$
    a. $1\dfrac{17}{24}x + 1\dfrac{5}{6}$ b. $-\dfrac{1}{24}x - 5\dfrac{1}{6}$

    c. $-\dfrac{1}{24}x + 1\dfrac{5}{6}$ d. $-\dfrac{1}{24}x + 5\dfrac{1}{6}$

4. _____

Solve each of the following equations.

5. $\dfrac{1}{7} + x = -4$
    a. $-\dfrac{29}{7}$ b. $\dfrac{27}{7}$

    c. $\dfrac{29}{7}$ d. $-\dfrac{27}{7}$

5. _____

6. $6g + 5 - 2(g + 2) = -(-7g - 1)$

6. _____

    a. $\dfrac{2}{3}$ b. 0

    c. -10 d. $-\dfrac{8}{15}$

7. $\dfrac{1}{3}(15x - 6) = 20\left(\dfrac{2}{5}x - \dfrac{1}{4}\right) - 3x$

7. _____

    a. $\dfrac{7}{16}$ b. $\dfrac{1}{2}$

    c. all real numbers   d. no solution

8. $3(1 - 6h) = -2(3h + 1) + 9$

8. _____

    a. $-\dfrac{1}{3}$ b. $\dfrac{7}{6}$

    c. $\dfrac{1}{6}$ d. $-\dfrac{2}{3}$

9. $3a + 2[2 + 3(a - 1)] = 2(4a - 1) + a$

9. _____

    a. 1 b. 2

    c. all real numbers   d. no solution

10. $\dfrac{4x + 5}{3} = \dfrac{2x + 6}{5}$

10. _____

    a. $\dfrac{15}{2}$ b. $-\dfrac{1}{2}$

    c. $\dfrac{43}{14}$ d. $-\dfrac{43}{14}$

11. $0.15(x-2)-0.4(x+1)=0.1(25)$

    a. -1.8 b. -3.8

    c. -1.4 d. -12.8

11. _____

12. $-2\left[3-(7+2x)\right]-(5-x)=7x$

    a. -1.3 b. -2.5

    c. 1.5 d. 0.30

12. _____

13. $0.5x-1.2=0.35$

    a. -8 b. $\dfrac{24}{17}$

    c. 8 d. $-\dfrac{24}{17}$

13. _____

Solve each of the following applications.

14. Find the value of r if $C=113.04$, for the equation $C=2\pi r$. Use 3.14 for π.

    a. 20   b. 36

    c. 18   d. 24

14. _____

15. Three times the sum of a number and 2, decreased by five times a number is zero.

    a. 1   b. 3

    c. 0.50   d. -3

15. _____

16. A $54 shirt is on sale for 25% off. If the sales tax is 8%, what would the total of the purchase?

    a. $15.75   b. $13.50

    c. $43.74   d. $40.50

16. _____

17.   I have invested a sum of money into an account that gives 5.5% interest. I invested $250 more in my account that gives 7% interest. If the total interest at the end of the year is $800, how much did I invest altogether?

    a. $6260   b. $6290

17. _____

    c. $12,770   d. $12,830

18.   There was an accident at the candy factory, and two types of candies were mixed together. There were 20 pounds of the chocolate chips that sell for $1.50 per pound and 40 pounds of the peanut butter chips that sell for $2.50 per pound. How much should the candy store sell the mixture so that the owners will not lose money?

    a. $5.00   b. $2.17

18. _____

    c. $ 4.23   d. $ 3.72

19.   Aaron has $6.25 in quarters and dimes. He has 10 more dimes than quarters. How many quarters does he have?

    a. 15   b. 5

19. _____

    c. 13   d. 30

Solve each of the following equations for the indicated variable.

20. $A = \dfrac{1}{2}bh$ for b.

    a. $\dfrac{2A}{h} = b$ b. $\dfrac{A}{2h} = b$

20. _____

    c. $\dfrac{Ah}{2} = b$ d. $\dfrac{2h}{A} = b$

21. $P = 2L + 2W$ for L.

    a. $P - W = L$ b. $\dfrac{P - W}{2} = L$

21. _____

    c. $P - 2W = L$ d. $\dfrac{P - 2W}{2} = L$

Solve each of the following. Write your answers in interval notation.

22. $2x > 6x - 24$

    a. $(6, \infty)$

    b. $[6, \infty]$

    c. $(-\infty, 6)$

    d. $(-\infty, 6]$

22. _____

23. $4(p - 4) \le 7p - 8$

    a. $\left(-\dfrac{8}{3}, \infty\right)$

    b. $\left[-\dfrac{8}{3}, \infty\right)$

    c. $\left(-\infty, -\dfrac{8}{3}\right)$

    d. $\left(-\infty, -\dfrac{8}{3}\right]$

23. _____

24. $-12 \le 6(x - 8) < 30$

    a. $(6, 13)$

    b. $[6, 13)$

    c. $(-6, 13)$

    d. $(-6, 13]$

24. _____

25. The three sides of a triangle are given to be three consecutive integers. If the perimeter is greater than 60 but less than 120, what must the requirements for the smallest side be?

    a. $(18, 38)$

    b. $(17, 37)$

    c. $(20, 40)$

    d. $(19, 39)$

25. _____

Martin-Gay *Beginning Algebra, Fifth Edition*

Chapter 3 Graphing
Section 3.1 Reading Graphs and the Rectangular Coordinate System

Learning Objectives
1. Read bar and line graphs.
2. Define the rectangular coordinate system and plot ordered pairs of numbers.
3. Graph paired data to create a scatter diagram.
4. Determine whether an ordered pair is a solution of an equation in two variables.
5. Find the missing coordinate of an ordered pair solution, given one coordinate of the pair.

Vocabulary.
Use the choices to complete each statement.

False	Four	Origin
Quadrants	Solutions	True
x-axis	x-coordinate	y-axis
y-coordinate	(0, 0)	(10, 10)

1. The rectangular coordinate system is created with a vertical line called the _____ and a horizontal line called the _____.

2. The rectangular coordinate system is separated into _____ regions that we refer to as _____.

3. A special point found at the intersection of the two axes is called the _____. Its ordered pair is _____.

4. True or false. Any ordered pair represents one point in the plane. _____.

5. In an ordered pair, the first number is the _____ and the second value in the pair is the _____.

6. The _____ of an equation in two variables is the ordered pair that makes the equation a true statement.

Objective 1

The following bar graph contains the percent of public school instructional rooms with internet access. (Source: National Center for Education Statistics) Use this bar graph to answer the following questions.

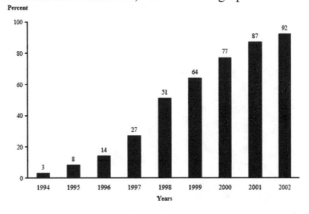

7. In between what years was there the greatest increase?

7. _____

8. What was that difference?

8. _____

9. By looking at this bar graph, can you see a relationship between the two given variables? What do you see?

9. _____

The following line graph contains the number of "unruly passengers" according to FAA regulations. Use this line graph to complete the following questions.

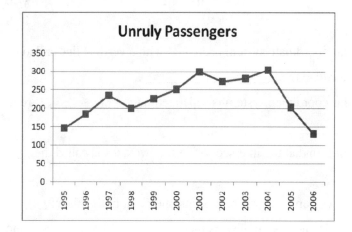

10. In what year were there the least unruly passengers?

10. _____

11. About how many unruly passengers was there in 2000?

11. _____

12. Is there a trend between the years and the number of unruly passengers? Does the year affect the number of unruly passengers?

12. _____

Objective 2

Plot each ordered pair. State in which quadrant or axis each point lies.

13.   A. $(5,7)$

          B. $(-4,9)$

          C. $(9,-5)$

          D. $(-3.5,0)$

          E. $\left(0,\dfrac{12}{5}\right)$

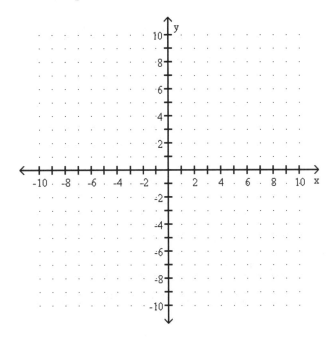

Objective 3

14. The table shows the percentage of 11th graders that said they used their computers to learn something. (Source: National Center of Education Statistics.)

Year	Percentage
1984	54.6
1988	65.3
1990	64.5
1992	72.3
1994	70.7
1996	80.2

     a. Write each paired data as an ordered pair of the form (year, percentage)

                                           14 a. _____

     b. Draw a grid and create a scatter diagram of the paired data.

Objective 4

Determine whether each ordered pair is a solution of the given linear equation.

 15. $2x + y = 7$ $(3,1)\ (7,0)\ (0,7)$

15. _____

16. $-3y + 4x = -12$ $(3,0)\ (2,-1)\left(-2,\dfrac{4}{3}\right)$

16. _____

Objective 5

Complete each ordered pair so that it is a solution of the given linear equation.

 17. $x - 4y = 4$ $(\ \ ,-2)\ (4,\ \ )$

17. _____

18. $2y = \dfrac{1}{3}x - 4$ $(-9,\ \ )(\ \ ,3)$

18. _____

Complete the table of ordered pairs for each linear equation.

19. $y = -4x$

x	y
0	
	-8
- 3	

Martin-Gay *Beginning Algebra, Fifth Edition*

20. $y = \dfrac{1}{2}x - 5$

x	y
0	
	0
-6	

Concept Extensions

21. Three vertices of a rectangle are $(3.2, 5.4)$, $(3.2, -2.2)$, and $(-1.6, 5.4)$

   a. Find the coordinate of the fourth vertex.

   21 a. _____

   b. Find the perimeter of the rectangle.

   21 b. _____

   c. Find the area of the rectangle.

   21 c. _____

Section 3.2 Graphing Linear Equation

Learning Objectives
1. Identify linear equations.
2. Graph a linear equation by finding and plotting ordered pair solutions.

Objective 1

Determine whether the given equation is a linear equation in two variables.

1. $1.4x - 2.8y = 6.4$ 1. _____

2. $y = -3$ 2. _____

3. $\dfrac{2}{3}x - \dfrac{1}{7}y = -2$ 3. _____

4. $x + y^4 = -1$ 4. _____

Objective 2

For each equation, find three ordered pair solutions by completing the table. Then use the ordered pairs to graph the equation.

5. $x + y = 5$

x	y
0	
	0
2	

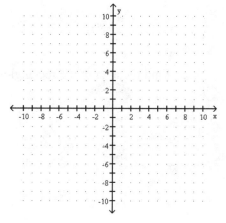

6. $y = \dfrac{1}{3}x - 4$

x	y
0	
6	
-9	

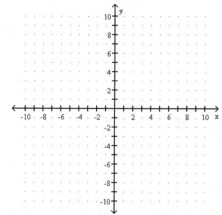

                                                        Martin-Gay *Beginning Algebra, Fifth Edition*

7. $y = -3x - 2$

x	y
0	
1	
2	

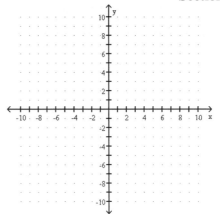

Graph each linear equation.

8. $x + y = -3$

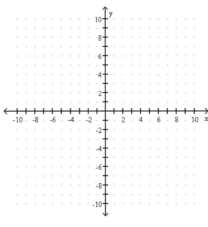

 9. $x - 2y = 6$

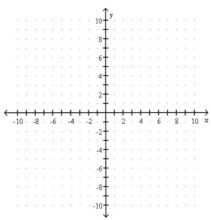

10. $y = -2$

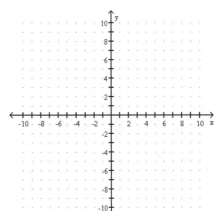

Martin-Gay *Beginning Algebra, Fifth Edition*   81

11. $2x - 4y = -8$

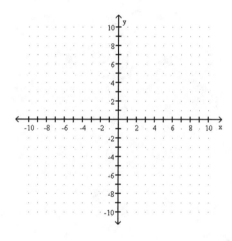

12. $y = 0.25x - 4$

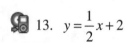 13. $y = \dfrac{1}{2}x + 2$

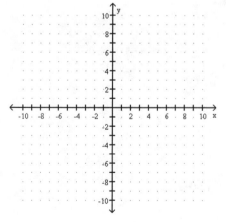

14. Graph each pair of linear equations on the same set of axes. Discuss how the graphs are similar and how they are different.

$y = 3x$ $y = 3x - 4$

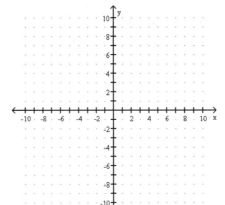

15. $y = -2x + 3$ $y = 2x + 3$

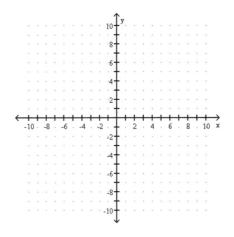

Solve.

16. The cost for a company to produce videos is $1.75 per video plus $1200 to run the equipment each month. The equation for costs of each month is $C = 1.75V + 1200$. How much would it cost the company to make 1000 videos?

16. _____

17. If the company from problem 16, decides to charge $19 for each video, the equation for profit is $P = 19V - 1200$. How much profit would the company make if they made 1000 videos?

17. _____

Concept Extensions

18. Take the company from problem 16 and 17. Given the two equations. $P = 19V - 1200$ and $C = 1.75V + 1200$. A break-even point for a company is the number of units that needs to be sold in order for the company to neither make a profit nor lose money. A break-even point is where the profit will equal the cost. For this company is there a break even point? Round your answer up to the next whole number of videos.

18. _____

Section 3.3 Intercepts

Learning Objectives
1. Identify intercepts of a graph.
2. Graph a linear equation by finding and plotting the intercepts.
3. Identify and graph vertical and horizontal lines.

Vocabulary.
Use the choices to complete each statement.

x	y	Horizontal
Linear	**Standard**	**Vertical**
x-intercept	**y-intercept**	

1. The graph $y = 2$ is a graph of a _____ line.

2. A point on the graph where x = 0 is called the _____.

3. The _____ form is $Ax + By = C$.

4. The point on the graph where it crosses the x-axis is called the _____.

5. The graph $x = -0.5$ is a graph of a _____ line.

6. Any equation that can be written as $Ax + By = C$ is called a _____ equation in two

   variables.

Objective 1

Identify the x- and the y-intercepts.

1.
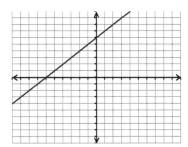

1.

2.
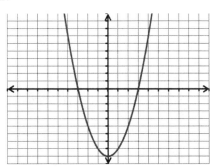

2.

Objective 2

Graph the equations by finding and plotting intercepts.

3. $2x - 3y = 12$

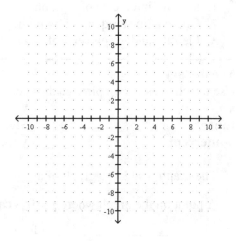

4. $-x + 2y = 6$

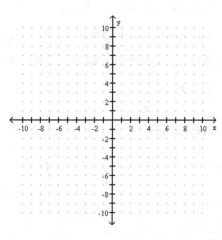

5. $y = \frac{1}{2}x + 3$

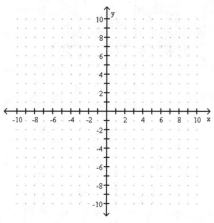

                                                     Martin-Gay *Beginning Algebra, Fifth Edition*

6. $y = -0.25x + 2$

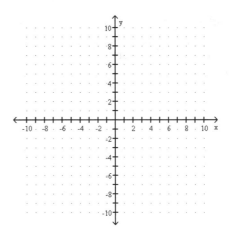

7. $y = -2x$

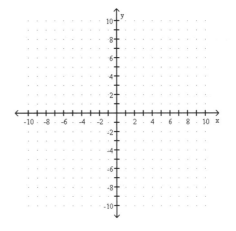

8. $4x + 3y = 5$

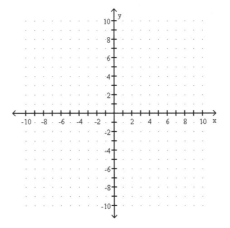

Objective 3

Graph each linear equation.

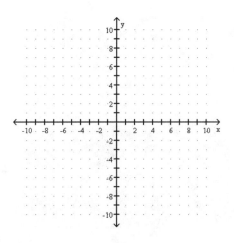

9. $y = 5$

10. $y + 3 = 7$

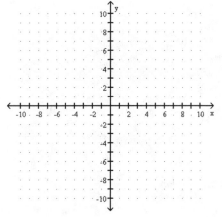

11. $x = 0$

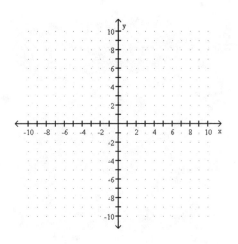

Martin-Gay *Beginning Algebra, Fifth Edition*

12. $x + 4 = 0$

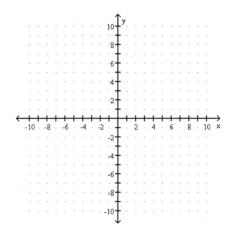

Concept Extensions

13. Do all graphs have x- and y-intercepts?

14. Are $x + y = -4$ and $-x - y = 4$ the same line? How do you know this?

Section 3.4 Slope and Rate of Change

Learning Objectives
1. Find the slope of a line given two points of the line.
2. Find the slope of a line given its equation.
3. Find the slopes of horizontal and vertical lines.
4. Compare the slopes of parallel and perpendicular lines.
5. Slope as a rate of change.

Vocabulary.
Use the choices to complete each statement.

b	*m*	**0**
Negative	**Parallel**	**Perpendicular**
Positive	**Slope**	**Undefined**

1. In the equation $y = mx + b$, _____ is the slope and _____ is the y-intercept.

2. If the value of y increases as the value of x increases, the line has a _____ slope.

3. A vertical line has a slope of _____.

4. The _____ of a line is the ratio of vertical change to its horizontal change.

5. Two lines are said to be _____ if they have the same slope, but different intercepts.

6. A horizontal line has a slope of _____.

Objective 1

Find the slope of the line that passes through the given points.

 7. $(-1,5)$ and $(6,-2)$

7. _____

8. $(2,-3)$ and $(4,-1)$

8. _____

9. $(0,-7)$ and $(0,3)$

9. _____

10. $(-4,9)$ and $(-11,18)$

10. _____

Objective 2

Find the slope of the given lines.

11. $y = \dfrac{2}{3}x - 7$

11. _____

 12. $2x + y = 7$

12. _____

13. $4x - 5y = -9$

13. _____

Objective 3

Find the slope of the given line.

14. $x = -7$

14. _____

15. $y = 24$

15. _____

Objective 4

Determine whether each pair of lines is parallel, perpendicular or neither.

 16.

16. _____

$y = \dfrac{2}{9}x + 3$

$y = -\dfrac{2}{9}$

17.

17. _____

$3x + 4y = 7$

$6x - 8y = -3$

18.   18. _____

$$y = -\frac{4}{9}x - 2$$
$$9x - 4y = -3$$

Objective 5

19. The pitch of a roof rises 6.5 feet over a horizontal distance of 13 feet. What is the slope of the roof?

19. _____

20. As a car drives 20 miles, the elevation of the road decreases by 5 feet. What is the grade of the road?

20. _____

Concept Extensions

21. The percentage of women who had an infant under the age of 1 and were in the labor force in 1976 was 31%. In 1998, the percentage was 59%.

    a. Rewrite this data as two ordered pairs, where x is the number of years since 1976.

21 a. _____

    b. Find the equation of the line in slope intercept form.

21 b. _____

    c. If the trend stays the same, what would the percentage be in 2007.

21 c. _____

22. What year would the percentage be 100%? Round your answer to the next whole year.

22. _____

Section 3.5   Equations of Lines

Learning Objectives
1. Use the slope-intercept form to write an equation of the line.
2. Use the slope-intercept form to graph a linear equation.
3. Use the point-slope form to find an equation of a line given its slope and a point of the line.
4. Use the point-slope form to find an equation of a line given two points of the line.
5. Find the equations of vertical and horizontal lines.
6. Use the point-slope form to solve problems.

Vocabulary.
Use the choices to complete each statement.

Horizontal   **Point-slope**   **Slope-intercept**
Standard   **Vertical**

1. The form $y - y_1 = m(x - x_1)$ is known as _____ form.

2. If the equation of a line is in the form y = constant, it is a _____ line.

3. The form $y = mx + b$ is referred to as _____ form.

4. An example of _____ form is $3x + 4y = 9$.

5. An equation of the form x = constant, is a _____ line.

Objective 1

Write the equation of the line with each given slope, m, and y-intercept $(0, b)$. Write the answer in slope-intercept form.

6. $m = 2, \ b = 4$

6. _____

7. $m = \dfrac{2}{3}, \ b = -2$

7. _____

8. $m = 0, \ b = -1$

8. _____

Objective 2

Use the slope-intercept form to graph each equation.

9. $y = 2x - 5$

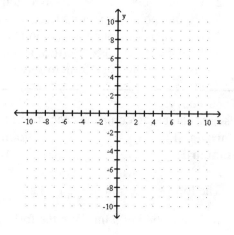

10. $y = -\dfrac{1}{4}x + 5$

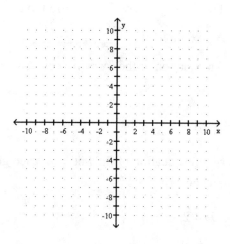

11. $4x - 7y = -14$

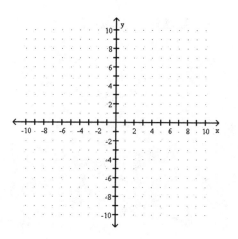

Objective 3

Find the equation of each line with the given slope that passes through the given point. Write the equation in the standard form.

12. $m = 3$ $(4, 3)$

12. _____

 13. $m = -8$ $(-1, -5)$

13. _____

14. $m = -\dfrac{1}{4}$ $(-2, -4)$

14. _____

Objective 4

Find an equation of the line passing through each pair of points. Write your answer in standard form.

15. $(1, -2)$ and $(-3, 4)$

15. _____

16. $(0, -4)$ and $(7, 2)$

16. _____

17. $(-4, 2)$ and $(0, -5)$

17. _____

Objective 5

Find the equation of each line.

18. Horizontal line through $(-12,18)$.

18. _____

19. Vertical line through $(24,-16)$

19. _____

20. Parallel to the x-axis, through $(2,-3)$

20. _____

21. Perpendicular to $y = 2$, through $(-7,9)$

21. _____

Objective 6

Solve. Assume the exercise describes a linear relationship. Write the equation in slope intercept form.

 22. A rock is dropped from the top of a 400-foot cliff. After 1 second, the rock is traveling 32 feet per second. After 3 seconds, the rock is traveling 96 feet per second.

22. _____

Concept Extensions

23. Suppose there is a line with a slope of $\dfrac{2}{3}$ and contains the point $(4,6)$. Are the points $(10,7)$ and $(-8,-2)$ also on the line? What is the equation of the line in slope-intercept form?

23. _____

Section 3.6 Functions

Learning Objectives
1. Identify relations, domains, and ranges.
2. Identify functions.
3. Use the vertical line test.
4. Use function notation.

Vocabulary.
Use the choices to complete each statement.

Domain	Function	Horizontal
Range	Relation	Vertical

1. Every linear equation is a function except for the equation that represents a _____ line.

2. A _____ is the set of ordered pairs.

3. A _____ is a set of ordered pairs, that for every x-value there is only one y-value.

4. The _____ of a relation is the set of all its x-coordinates.

5. The _____ of a relation is the set of al its y-coordinates.

Objective 1

Find the domain and range of each relation.

6. $\{(1,3)(-3,8)(0,-5)(3,-4)\}$

6. _____

7. $\{(2,-7)(8,-1)(-7,0)(1,-2)\}$

7. _____

8. $\{(22,-33)(-13,19)(10,36)(-18,-17)\}$

8. _____

Objective 2

Determine if the given relation is a function.

9. $\{(2,-4)(-3,18)(10,18)(3,4)\}$

9. _____

Martin-Gay *Beginning Algebra, Fifth Edition*

97

10. $\{(1,-2)(4,4)(1,-5)(2,-3)\}$

10. _____

Determine whether the given graph is a function.

11.

11. _____

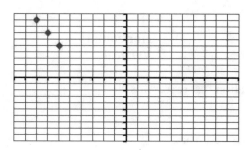

12.

12. _____

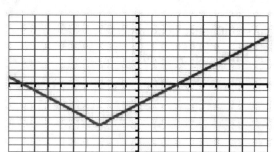

13.

13. _____

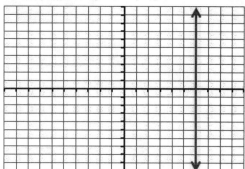

Determine whether the equation describes a function.

14. $y = 4x + 2$

14. _____

15. $\frac{1}{3}x - 1.7y = 0$ 15. _____

16. $x = y^2 - 7$ 16. _____

17. $x^2 - y = 21$ 17. _____

Objective 4

 18. Find $f(-2)$, $f(0)$, and, $f(3)$ for the given function. $f(x) = x^2 + 2$

18. _____

19. Find $g(-1)$, $g(2)$, and, $f(0)$ for the given function. Write each solution as an ordered pair.

$g(x) = x^3 + 2x^2 - 3x + 6$

19. _____

Find the domain and range of each relation graphed.

20.   20. _____

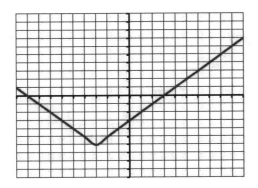

21.   21. _____

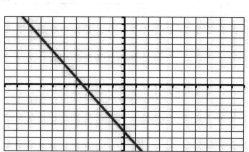

Find the domain of the given function.

22. $f(x) = x + 7$ 22. _____

23. $h(x) = \dfrac{4}{3 - 2x}$ 23. _____

24. $g(x) = |2x - 3|$ 24. _____

Concept Extension

25. Forensic scientists use the function $H(x) = 2.59x + 47.24$ to estimate the height of a woman in centimeters given the length x of her femur bone.

    a. Estimate the height of a woman whose femur measures 46 cm.

                                           25 a. _____

    b. Estimate the height of a woman whose femur measures 39 cm.

                                           25 b. _____

Chapter 3 Vocabulary

Vocabulary Word	Definition	Example
Rectangular coordinate system	A plane that contains a vertical (y) and horizontal (x) axes. The intersection of the axes is called the origin.	
Solution	An ordered pair the makes the equation true.	$y = 3x + 2$ Solution $(-2, -4)$ $-4 = 3(-2) + 2$
Linear Equation in Two Variable	An equation that can be written in standard form: $Ax + By = C$	$5x - 4y = 12$ $y = 2x - 8$
Intercept	A point on the graph where it hits a specific axis.	$y = 2x + 8$ Hits the y axis at 8.
Slope (m)	Rate of change. Steepness of the graph.	$\dfrac{y_2 - y_1}{x_2 - x_1}$
Slope-Intercept Form.	$y = mx + b$ where m is the slope and b is the y-intercept.	$y = 4x - 9$
Point-Slope Form	$y - y_1 = m(x - x_1)$ where m is the slope and the point (x_1, y_1)	Given m = 2 and $(1,3)$ then $y - 3 = 2(x - 1)$
Relation	Set of ordered pairs. A function is a special relation that the x values do not repeat. For every x-value there is only one y-value.	$\{(1,2)(2,3)(4,5)\}$ Is a function as well.
Domain	Set of all the x-values in a relation.	For the relation above: D={1, 2, 4}
Range	Set of all the y-values in a relation.	For the relation above: R= {2, 3, 5}
$f(x)$	Function notation. A function of x.	$f(x) = 3x + 2$

Chapter 3 Practice Test A

Complete the ordered pair so that it is a solution to the given linear equation.

1. $y = -3x + 4$ $(-3, \ )(\ , 4)(7, \ )$

1. _____

State in which quadrant, if any, the given point lies.

2. $(-7, -8)$

2. _____

Determine whether the ordered pair is a solution of the given linear equation.

3. $4x - 3y = 8$ $(5, -4)$

3. _____

Determine whether the equation is a linear equation in two variables.

4. $x^2 = y - 19$

4. _____

Graph the given linear equation.

5. $4y = 2x - 8$

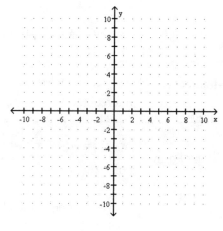

6. $y = -2x + 6$

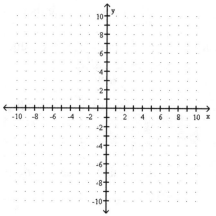

Martin-Gay *Beginning Algebra, Fifth Edition*

Identify the intercepts.

7.

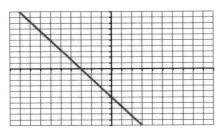

Graph the linear equation by finding the x- and y-intercepts.

8. $6y - 9x = 18$

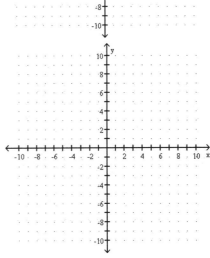

9. $x = -\dfrac{1}{2}y$

Graph the linear equation.

10. $x = -4$

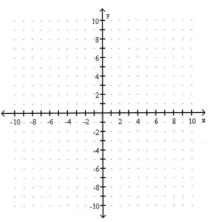

Find the slope of the line that goes through the given points.

11. $(-3,5)$ and $(5,7)$

11. _____

Find the slope of the line.

12.

12. _____

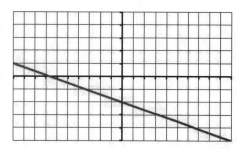

13. $3x - 7y = 9$

13. _____

Determine whether the given pair of lines are parallel, perpendicular, or neither.

14.

$$y = \frac{4}{5}x - 9$$

$$5x - 4y = 12$$

14. _____

Determine the slope and y-intercept of the given linear equation.

15. $10x - 18y = 24$

15. _____

Write the equation of the line in slope-intercept form.

16. $m = \frac{4}{7}$ $(0,-6)$

16. _____

Use the slope-intercept form the graph the equation.

17. $y = \dfrac{1}{3}x - 5$

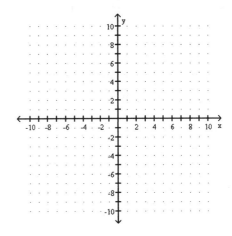

18. $5y - 4x = 15$

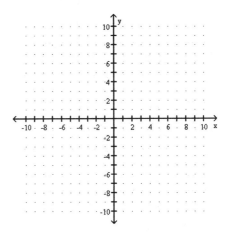

Write the equation of each line in standard form.

19. $m = -4$ $\qquad$ $(-5, -7)$

19. _____

20. $m = \dfrac{2}{5}$ $\qquad$ $(-3, 4)$

20. _____

21. A vertical line, through $(-14, 16)$.

21. _____

22. Through $(-33, 24)$, perpendicular to $x = 22$.   22. _____

Determine if the following is a function.

23. $\{(-1,5)(2,5)(-5,-5)\}$ 23. _____

24. $y = |x - 4|$ 24. _____

25. Given $h(x) = x^2 - 3x + 9$, find the indicated function values. Write the solutions as ordered pairs.

   $h(-4)$ $h(0)$ $h(3)$

                                                             25. _____

Chapter 3 Practice Test B

Complete the ordered pair so that it is a solution to the given linear equation.

1. $y = 5x - 8$ $(-4, \ )(\ ,12)(2, \ )$ 1. _____

     a. $(-4, 28)(4,12)(2,2)$ b. $(-4, -28)(4,12)(2,-2)$

     c. $(-4, 28)(4,12)(2,-2)$ d. $(-4, -28)(4,12)(2,2)$

State in which quadrant, if any, the given point lies.

2. $(3, -9)$ 2. _____

     a. I   b. II

     c. III   d. IV

Determine whether the ordered pair is a solution of the given linear equation.

3. $-3x - 9y = 6$ $(4, -2)$ 3. _____

     a. Yes   b. No

Determine whether the equation is a linear equation in two variables.

4. $5.7y - 4.9x = 13.7$ 4. _____

     a. Yes   b. No

Graph the given line.

5. $2x + 5y = 15$ 5. _____

   a..   b.

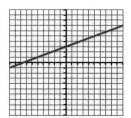   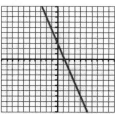

   c.   d.

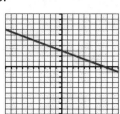   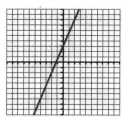

6. $y = -\dfrac{2}{3}x - 4$ 6.

   a.   b.

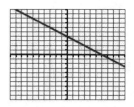   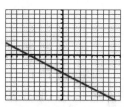

   c.   d.

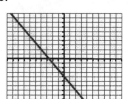   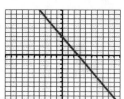

Determine whether the given pair of lines are parallel, perpendicular, or neither.

7.

   $3y = 6x - 4$

   $y = \dfrac{1}{2}x - 9$ 7. _____

   a. Parallel   b. Perpendicular

   c. Neither   d. Identify the intercepts.

8.

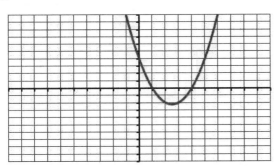

   a. $(0,2)(0,3)(4,0)$ b. $(2,3)(4,0)$

   c. $(2,0)(3,0)(4,0)$ d. $(1,0)(3,0)(0,4)$

Find the slope of the line that goes through the given points.

9. $(-1,6)$ and $(3,-4)$ 9. _____

   a. $\dfrac{2}{5}$ b. $-\dfrac{5}{2}$

   c. $\dfrac{5}{2}$ d. $-\dfrac{2}{5}$

Graph the linear equation by finding the x- and y-intercepts.

10. $9y - 4x = 18$ 10. _____

      a.   b.

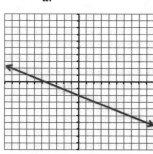   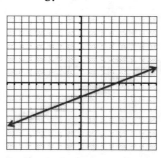

      c.   d

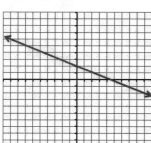   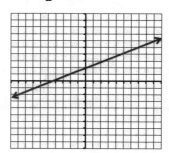

Write the equation of the line in slope-intercept form.

11. $m = \dfrac{3}{4}$ $(0,8)$ 11. _____

    a. $y = \dfrac{3}{4}x + 8$ b. $y = \dfrac{3}{4}x - 8$

    c. $y = 8x + \dfrac{3}{4}$ d. $4x - 3y = 8$

Graph the linear equation.

12   $y = -3$ 12. _____

      a.   b.

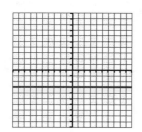   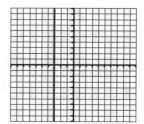

Graph the linear equation by finding the x- and y-intercepts.

13. $x = -4y$ 13. _____

a.   b.

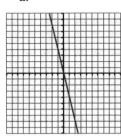   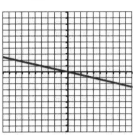

c.   d..

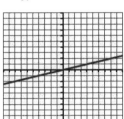   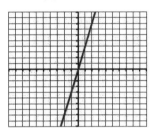

Find the slope of the line.

14.   14. _____

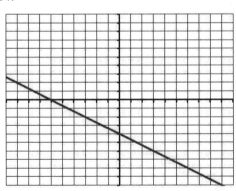

a. $\dfrac{2}{3}$ b. $\dfrac{3}{2}$

c. $-\dfrac{3}{2}$ d. $-\dfrac{2}{3}$

15. $-3x - 5y = 15$ 15. _____

    a. 3   b. $\dfrac{5}{3}$

    c. $-\dfrac{3}{5}$ d. $-\dfrac{5}{3}$

Determine the slope and y-intercept of the given linear equation.

16. $4x - 5y = 25$ 16. _____

    a. $m = \dfrac{5}{4}$ $b = 5$ b. $m = \dfrac{4}{5}$ $b = -5$

    c. $m = -\dfrac{4}{5}$ $b = -5$ d. $m = \dfrac{5}{4}$ $b = -5$

Write the equation of each line in standard form.

17. $m = 2$ $(1,3)$ 17. _____

    a. $2x - y = 1$ b. $2x - y = -1$

    c. $y = 2x - 1$ d. $-2x - y = 1$

18. A horizontal line, through $(-8,11)$ 18. _____

    a. $y = 11$ b. $x = 11$

    c. $y = -8$ d. $x = -8$

Use the slope-intercept form to graph the equation.

19   $y = -\dfrac{2}{5}x - 8$ 19. _____

a.   b.

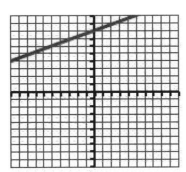   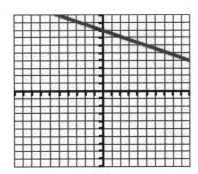

c.   d.

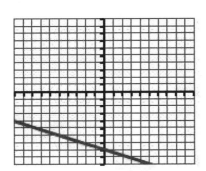   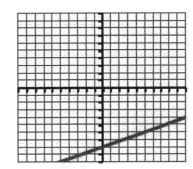

Determine if the following is a function.

20. $\{(2,3)(9,-3)(2,13)\}$ 20. _____

　　　　a. It is a function.   b. It is not a function.

21. $y = 4$ 21. _____

　　　　a. It is a function.   b. It is not a function.

22. Given $g(x) = |3x - 2|$, find the indicated function values. Write the solutions as ordered pairs.

    $g(-2)$ $g(0)$ $g(5)$

22. _____

    a. $(-2, -8)(0, -2)(5, 13)$ b. $(-2, 8)(0, 2)(5, 13)$

    c. $(8, -2)(2, 0)(5, 13)$ d. $(-2, -8)(0, -2)(5, 13)$

Use the slope-intercept form to graph the equation.

23. $7x + 4y = 28$ 23. _____

    a.   b.

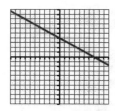   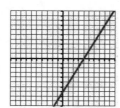

    c.   d.

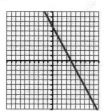   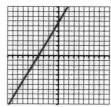

Write the equation of each line in standard form.

24. Through $(13, -14)$, parallel to $x = 2$.   24. _____

    a. $x = 13$ b. $y = 13$

    c. $x = -14$ d. $y = -14$

25. $m = -\dfrac{1}{3}$ $(-6, 1)$ 25. _____

       a. $x + 3y = -3$ b. $3x - y = 3$

       c. $-x + 3y = 9$ d. $x + 3y = 9$

Name: _____   Date: _____

Instructor: _____   Section: _____

Chapter 4 Solving Systems of Linear Equations and Inequalities
Section 4.1 Solving Systems of Linear Equations by Graphing

Learning Objectives
1. Determine if an ordered pair is a solution of a system of equations in two variables.
2. Solve a system of linear equations by graphing.
3. Without graphing, determine the number of solutions of a system.

Vocabulary.
Use the choices to complete each statement.

Consistent	Dependent	Inconsistent
Independent	Solution	System of linear equations

1. If the system of equations has two distinct lines, it is called a(n) _____ system.

2. A set of two or more linear equations is called a(n) _____.

3. If a system of equations has at least one or more solutions, it is called a(n) _____

   system.

4. When graphing a system of equations, if both lines turn out to be the same line, it is called a(n)

   _____ system.

5. Parallel lines make a(n) _____ system because the system will have no

   solution.

6. A _____ of a system of equations is an ordered pair that not only works in one of the

   equations, but all the equations in the system.

Objective 1
Determine whether each ordered pair is a solution of the system of linear equations.

7. $\begin{cases} x+y=16 \\ y-x=2 \end{cases}$

    a. $(7,9)$ 7 a. _____

    b. $(10,6)$ 7b. _____

 8. $\begin{cases} 3x-y=5 \\ x+2y=11 \end{cases}$

    a. $(3,4)$ 8a. _____

    b. $(0,-5)$ 8b. _____

116   Martin-Gay *Beginning Algebra, Fifth Edition*

9. $\begin{cases} 2y = -3x - 8 \\ 5x - 6y = -4 \end{cases}$

     a. $(4, 4)$

     b. $(-2, -1)$

9a. _____

9b. _____

Objective 2

Solve the system by graphing.

10. $\begin{cases} x = \dfrac{3}{4}y + 1 \\ 7x - 3y = -2 \end{cases}$

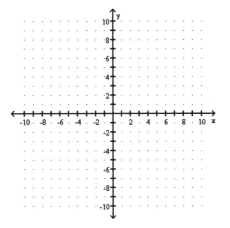

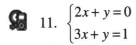 11. $\begin{cases} 2x + y = 0 \\ 3x + y = 1 \end{cases}$

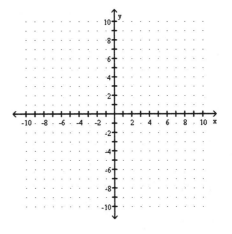

12. $\begin{cases} 2x - 4y = -6 \\ 3x - 6y = 10 \end{cases}$

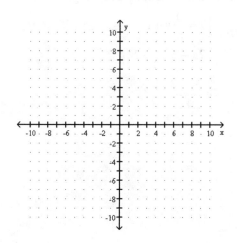

13. $\begin{cases} y = -2x + 3 \\ 6x + 3y = 9 \end{cases}$

Objective 3
Without graphing, decide if the graphs of the equations are identical lines, parallel lines, or lines that are intersecting. Also state how many solutions will the system have?

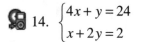 14. $\begin{cases} 4x + y = 24 \\ x + 2y = 2 \end{cases}$

14. _____

15. $\begin{cases} y = \dfrac{2}{3}x - \dfrac{4}{3} \\ 6x = 12 + 9y \end{cases}$

15. _____

16. $\begin{cases} 3x - 7y = -9 \\ 3x - 7y = 5 \end{cases}$

16. _____

17. $\begin{cases} \dfrac{2}{3}x + \dfrac{1}{5}y = 0 \\ \dfrac{3}{2}x - \dfrac{3}{10}y = -15 \end{cases}$

17. _____

Concept Extensions

18. Is $\left(\dfrac{37}{13}, -\dfrac{21}{13} \right)$ a solution of the system: $\begin{cases} 5x + 2y = 11 \\ 7x + 8y = 7 \end{cases}$?

18. _____

Section 4.2 Solving Systems of Linear Equations by Substitution

---
Learning Objectives
  1. Use the substitution method to solve a system of linear equations.

Objective 1
Solve each system of equations by the substitution method.

 1. $\begin{cases} x + y = 6 \\ y = -3x \end{cases}$

1. _____

2. $\begin{cases} 3x - y = 10 \\ y = -x + 2 \end{cases}$

2. _____

3. $\begin{cases} -x + 3y = -3 \\ 9y - 3x = -9 \end{cases}$

3. _____

4. $\begin{cases} y = -\dfrac{3}{4}x - 7 \\ y = -\dfrac{1}{2}x - 3 \end{cases}$

4. _____

5. $\begin{cases} -8x + 3y = 22 \\ 4x + 3y = -2 \end{cases}$

5. _____

6. $\begin{cases} 3x + 6y = 9 \\ 4x + 8y = 16 \end{cases}$

6. _____

7. $\begin{cases} 2x - 3y = -4 \\ 3x + 4y = -6 \end{cases}$

7. _____

8. $\begin{cases} 4x - 2y = 5 \\ x = \dfrac{1}{2}y + \dfrac{5}{4} \end{cases}$

8. _____

9. $\begin{cases} \dfrac{1}{3}x - y = 2 \\ x - 3y = 6 \end{cases}$

9. _____

10. $\begin{cases} 2x + y = 9 \\ 7x + 4y = 1 \end{cases}$

10. _____

11. $\begin{cases} -2y = 1 + 7x \\ 4x + 5y = 2 \end{cases}$

11. _____

12. $\begin{cases} 2x + 3y = 6 \\ y = 3x - 5 \end{cases}$

12. _____

13. $\begin{cases} 3x - 6y = 10 \\ 2x - 4y = -6 \end{cases}$

13. _____

14. $\begin{cases} y = 22 - x \\ 0.05x + 0.10y = 1.70 \end{cases}$

14. _____

15. $\begin{cases} 2x + 6y = -28 \\ 3x + 9y = -27 \end{cases}$

15. _____

16. $\begin{cases} 7x - 6y = -1 \\ x = 2y - 1 \end{cases}$

16. _____

17. $\begin{cases} 4x + 5y = 2 \\ 16x - 15y = 1 \end{cases}$

17. _____

Martin-Gay *Beginning Algebra, Fifth Edition*

Concept Extensions

18.

$$\begin{cases} \dfrac{x+2}{3} = \dfrac{3-y}{2} \\ \dfrac{x+3}{2} = \dfrac{2-y}{3} \end{cases}$$

18. _____

19. $\begin{cases} 2(x-1) - 3(y+2) = 30 \\ 3(x+2) + 2(y-1) = -4 \end{cases}$

19. _____

Section 4.3 Solving Systems of Linear Equations by Addition

Learning Objectives
   1. Use the addition method to solve a system of linear equations

---

Objective 1
Solve each system of equations by the addition method.

   1. $\begin{cases} x - 2y = 8 \\ -x + 5y = -17 \end{cases}$

1. _____

2. $\begin{cases} 2x + 9y = 2 \\ x - 2y = -12 \end{cases}$

2. _____

3. $\begin{cases} 5x + 2y = -4 \\ 5x - 3y = 6 \end{cases}$

3. _____

4. $\begin{cases} x + y = 48 \\ 12x + 14y = 628 \end{cases}$

4. _____

5. $\begin{cases} 10x - 2y = 2 \\ 5x - y = 1 \end{cases}$

5. _____

6. $\begin{cases} x + 2y = 0 \\ 2x - y = 0 \end{cases}$

6. _____

7. $\begin{cases} 2x = 3(y - 2) \\ 2(x + 4) = 3y \end{cases}$

7. _____

8. $\begin{cases} 8x - 4y = 18 \\ 3x - 8 = 2y \end{cases}$

8. _____

9. $\begin{cases} \dfrac{x}{3} - y = 2 \\ -\dfrac{x}{2} + \dfrac{3y}{2} = -3 \end{cases}$

9. _____

10. $\begin{cases} \dfrac{1}{2}x - \dfrac{1}{4}y = 1 \\ \dfrac{1}{3}x + y = 3 \end{cases}$

10. _____

11. $\begin{cases} \dfrac{7}{12}x - \dfrac{1}{2}y = \dfrac{1}{6} \\ \dfrac{2}{5}x - \dfrac{1}{3}y = \dfrac{11}{15} \end{cases}$

11. _____

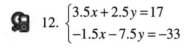 12. $\begin{cases} 3.5x + 2.5y = 17 \\ -1.5x - 7.5y = -33 \end{cases}$

12. _____

13. $\begin{cases} 3x - 2.5y = 0 \\ 2.5x - 3y = -1 \end{cases}$

13. _____

14. $\begin{cases} 0.1x + 0.06y = 9 \\ 0.09x + 0.5y = 52.7 \end{cases}$

14. _____

15. $\begin{cases} 3x + 2y = 3 \\ 4x - 3y = -13 \end{cases}$

15. _____

16. $\begin{cases} 0.07x + 0.3y = 6.70 \\ 7x + 30 = 67 \end{cases}$

16. _____

17. $\begin{cases} 4x + 7y = 2 \\ 9x - 2y = 1 \end{cases}$

17. _____

Concept Extension

18. $\begin{cases} \dfrac{x-y}{2} - \dfrac{2x-y}{3} = -\dfrac{1}{4} \\ \dfrac{2x+y}{3} + \dfrac{x+y}{2} = \dfrac{17}{6} \end{cases}$

18. _____

19. If possible find a solution to this system: $\begin{cases} x + y = 5 \\ x - y = -3 \\ 2x - y = -2 \end{cases}$

19. _____

Section 4.4 Systems of Linear Equations and Problem Solving

Learning Objectives
1. Use a system of equations to solve problems.

Objective 1

Solve.
1. Two numbers total 83 and have a difference of 17. Find the two numbers.

1. _____

2. The sum of two numbers is -16, and their difference is 8. Find the numbers.

2. _____

3. The width of a rectangle is 15 feet shorter than the length. If the perimeter is 82 feet. What are the dimensions of the rectangle.

3. _____

4. A plumber needs to cut a 25 foot pipe into two pieces so that one piece is 5 feet shorter than the other. How long would the longer piece?

4. _____

5. Twice one integer plus another integer is 21. If the first integer plus 3 times the second is 33, find the integers.

5. _____

6. A farmer raises five times as many chickens as goats. If he has 168 animals altogether, how many goats does the farmer have?

6. _____

7. Students can buy tickets to a baseball game for $1. The admission for nonstudents is $2. If 350 tickets are sold and the school earned $450, how many nonstudent tickets were sold?

7. _____

8. If the numerator of a fraction is increased by 5 and the denominator is decreased by 1, the result is the fraction $\frac{8}{3}$. If the numerator of that original fraction is doubled and the denominator is increased by 7, the new result is the fraction $\frac{6}{11}$. What is the original fraction?

8. _____

9. Ann Marie Jones has been pricing Amtrak train fares for a group trip to New York. Three adults and four children must pay $159. Two adults and three children must pay $112. Find the price of an adult's ticket, and find the price of a child's ticket.

9. _____

10. Sherry has 35 coins consisting of dimes and nickels. If the value of her coins is $3.30, then how many dimes does she have?

10. _____

11. George averaged 45 mph on the first leg of his trip and 49 mph on the second leg. If his total trip was 237 miles in 5 hours, then how far did he drive on his first leg?

11. _____

12. Recall that two angles are complementary if the sum of their measures is 90°. Find the measure of two complementary angles if one angle is twice the other.

12. _____

13. How many liters of 50% alcohol solution and 20% alcohol solution must be mixed to obtain 18 liters of 30% solution?

13. _____

14. A total of $11,000 was invested. Part of it at 4% , and the rest at 7% interest rate. If the investments ear $680 per year, how much was invested at each rate?

14. _____

15. Jean has $30,000, part of which she invests at 9% interest and the rest at 7%. If her income from the 9% is $820 less than that from the 7% investment, how much did Jean invest at 7% interest rate?

15. _____

16. A merchant wants to mix the peanuts with cashews. The peanuts are sold at $3 per pound and the cashews are sold at $6 per pound. The merchant wants to make a 48 pound mix of nuts to sell for $4 per pound. How many pounds of each should the merchant use?

16. _____

17. The length of a room is two feet less than twice its width. If its perimeter is 86, what is the area of the room?

17. _____

Concept Extensions.

18. George has $3.75 in nickels, dimes and quarters. If he had twice as many dimes, half as many nickels, and the same amount of quarters he would have $5.50. If he has 35 coins altogether, how many of each does he have?

18. _____

Section 4.5 Graphing Linear Inequalities

Learning Objectives
1. Graph a linear inequality in two variables.

Vocabulary.
Use the choices to complete each statement.

True	False
Boundary Line	**Linear Inequality in One Variable.**

1. A _____ will divide a plane into two regions called half-planes.

2. $4x + 5y < 12$ is an example of a _____.

3. True or False. The boundary line of the inequality $3x + 2y \geq -5$ is dashed. _____

4. True or False. The shading for the inequality $2x - 3y > 8$ is below the boundary line. _____

5. A _____ of a linear inequality is an ordered pair, that makes the inequality true.

Objective 1

Graph each linear inequality.

6. $x + y < 6$

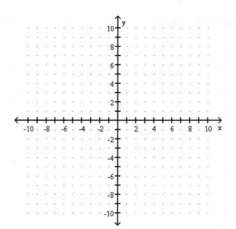

7. $x - 4y > 5$

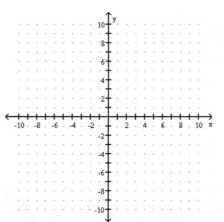

8. $2x + 7y > 5$

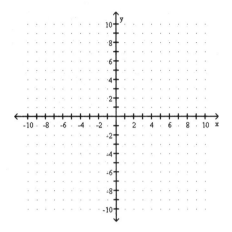

9. $4x - y \geq 6$

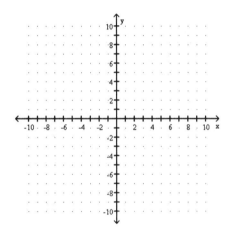

10. $-2x - 4y < -8$

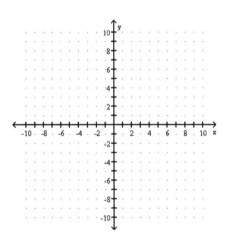

11. $y \geq 2x$

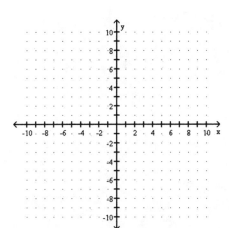

12. $y \leq -4$

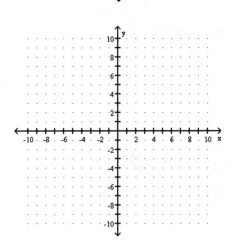

13. $x > -5$

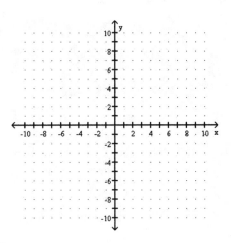

Martin-Gay *Beginning Algebra, Fifth Edition*

14. $x < 6y$

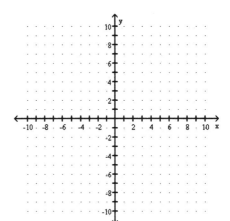

15. $-x < 3y + 9$

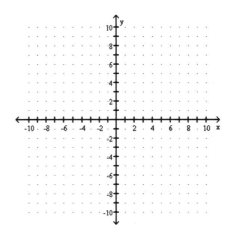

16. $\frac{1}{2}x - \frac{3}{8}y > \frac{5}{6}$

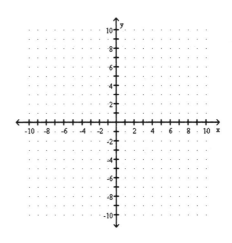

17. $0.5y - 0.3x < 2.5$

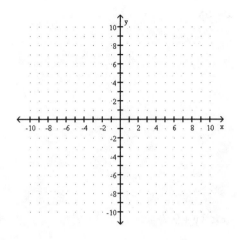

Concept Extensions

18. A local theater holds a maximum of 200 people. For a fundraiser, a youth group will be showing a play. The price for student tickets are $2.00 and the price for nonstudents tickets was set at $3.00 each. The group needs to sell at least $80 worth of ticket to make any profit. Create a graph using only the first quadrant to show all the possible combinations of ticket sales that will allow the youth group to make money.

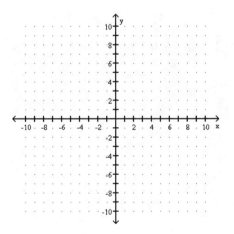

Section 4.6 Systems of Linear Inequalities

Learning Objectives
1. Solve a system of linear inequalities.

Objective 4.6.1

Solve the system of linear inequalities.

1. $\begin{cases} y \le \dfrac{1}{2}x + 3 \\ y > \dfrac{1}{2}x - 1 \end{cases}$

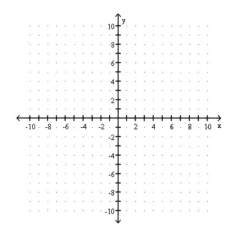

2. $\begin{cases} x - y \ge 4 \\ x + y < 3 \end{cases}$

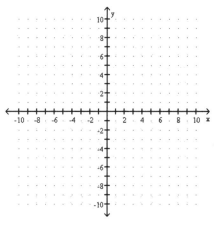

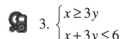 3. $\begin{cases} x \ge 3y \\ x + 3y \le 6 \end{cases}$

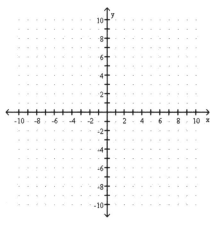

4. $\begin{cases} 3x - 4y > 12 \\ -3x - 2y \leq 6 \end{cases}$

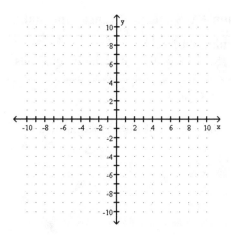

5. $\begin{cases} 2x + y < 3 \\ x - y < 2 \end{cases}$

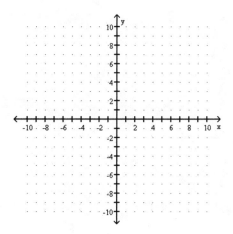

 6. $\begin{cases} y \geq 1 \\ x < -3 \end{cases}$

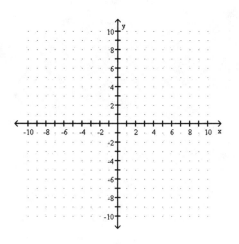

7. $\begin{cases} 4x - 5y \geq 0 \\ y \geq 2 \end{cases}$

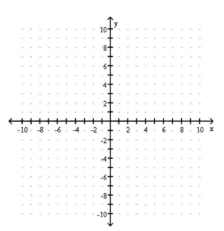

8. $\begin{cases} y < 3x \\ y > -2x \end{cases}$

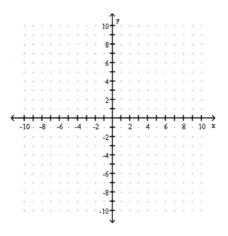

9. $\begin{cases} 4x - 2y > 8 \\ 3x - 2y < 6 \end{cases}$

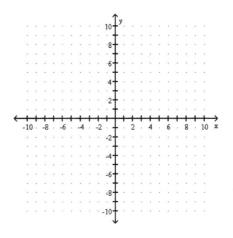

10. $\begin{cases} y > 2x - 4 \\ y \le 2x - 2 \end{cases}$

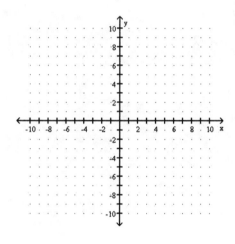

11. $\begin{cases} y > -4 \\ x < 5 \end{cases}$

12. $\begin{cases} x - 7y \ge 0 \\ x + y < 0 \end{cases}$

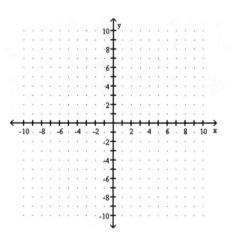

13. $\begin{cases} 3x - 6y < 12 \\ x - y \geq 3 \end{cases}$

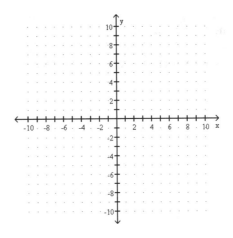

14. $\begin{cases} 6x + 4y < 4 \\ -2x - 4y > 8 \end{cases}$

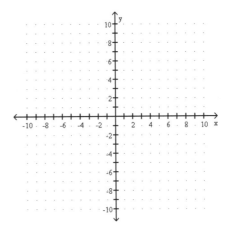

15. $\begin{cases} y \leq -\dfrac{2}{5}x + 2 \\ x > 3 - y \end{cases}$

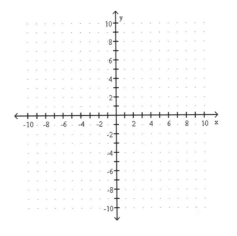

Concept Extension

16. Graph this system of linear equations.

$$\begin{cases} x \le 1 \\ y < 7 \\ 5x - y < -8 \\ 3x - 4y < -12 \end{cases}$$

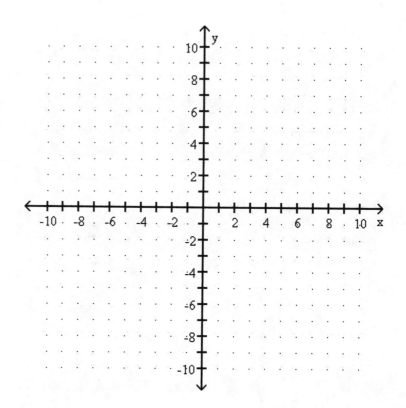

Chapter 4 Vocabulary

Vocabulary Word	Definition	Example
Solution	An ordered pair that makes all equations in the system true. An ordered pair that makes a linear inequality true.	$(1, 3)$ is a solution of $\begin{cases} y = x + 2 \\ x + y = 4 \end{cases}$ And it is a solution of $y > x + 1$
Consistent	A system of linear equations that has at least one solution.	
Inconsistent	A system of linear equations that has no solutions.	
Dependent	A system of linear equations that the graphs of each line are identical.	
Independent	A system of linear equations that the graphs of each line are unique.	

Linear Inequality in One Variable	$Ax + By < C$ $\quad$ $Ax + By > C$ $Ax + By \leq C$ $\quad$ $Ax + By \geq C$	$-3x + 4y < 17$
Substitution Method	Method to solve system of equations. Rewrite one of the equations to isolate a variable. Plug that expression into the other equation. Solve for remaining variable. Plug answer back into an equation to get the other variable.	$\begin{cases} x - y = 2 \\ x + y = 3 \end{cases}$ $x = 2 + y$ $2 + y + y = 3$ $2 + 2y = 3$ $2y = 1$ $y = \frac{1}{2}$ $x = 2 + \frac{1}{2} = \frac{5}{2}$
Elimination Method	Method to solve system of equations. Multiply equation(s) to make the coefficients of one of the variables opposites. Add the equations, this eliminates a variable. Solve for remaining variable. Plug answer back into an equation to get the other variable.	$\begin{cases} x - 2y = 3 \\ (x + y = 3) \cdot 2 \end{cases}$ $2x + 2y = 6$ $x - 2y = 3$ $3x = 9$ $x = 3$ $3 + y = 3$ $y = 0$

Practice Test A

1. Is $(-2,-3)$ a solution of $\begin{cases} y = x+1 \\ 3x = 2y \end{cases}$

1. _____

2. Is $(1,5)$ a solution of $\begin{cases} 2x + y = 7 \\ 3x - 4y = 17 \end{cases}$

2. _____

3. Solve the system by graphing. $\begin{cases} x - 5y = 10 \\ -2x + y = 7 \end{cases}$

3. _____

4. Solve the system by graphing. $\begin{cases} -3x + 2y = 4 \\ y = 2x + 3 \end{cases}$

4. _____

5. Solve the system by the substitution method. $\begin{cases} x - y = 9 \\ 4y - 3x = 5 \end{cases}$

5. _____

6. Solve the system by the substitution method. $\begin{cases} \frac{1}{2}x - \frac{2}{3}y = 3 \\ 6x - 5y = 18 \end{cases}$

6. _____

7. Solve the system by the elimination method. $\begin{cases} 4x - 2y = -1 \\ 3x + 6y = 18 \end{cases}$

7. _____

8. Solve the system by the substitution method. $\begin{cases} 0.25x + 0.5y = .75 \\ 0.2x - 0.4y = -.5 \end{cases}$

8. _____

9. Solve the system $\begin{cases} 12x + 10y = 24 \\ 5y + 3x = 16 \end{cases}$

9. _____

10. Solve the system $\begin{cases} y = -3x - 4 \\ 2x + 6y = 8 \end{cases}$

10. _____

11. Solve the system $\begin{cases} x = 3y - 9 \\ 2x - 6y = -18 \end{cases}$

11. _____

12. Graph the inequality. $x + 4y < 5$

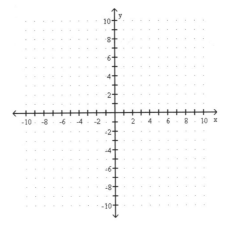

13. Graph the inequality. $2x - 7y > 14$

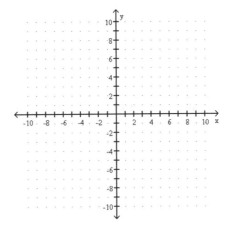

14. Graph the inequality. $-\dfrac{3}{4}x + \dfrac{3}{8}y \le -3$

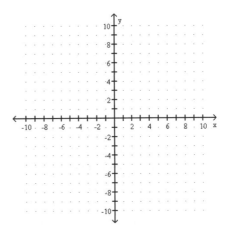

15. Graph the solutions of the system of linear inequality. $\begin{cases} 5x - 4y > 8 \\ -2x + 5y < 10 \end{cases}$

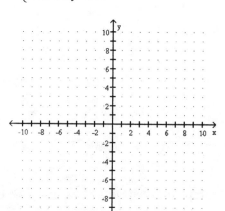

16. Graph the solutions of the system of linear inequality. $\begin{cases} x - y \geq 4 \\ x + y < -3 \end{cases}$

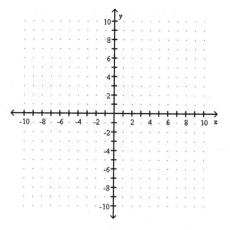

17. The sum of two numbers is seven. Their difference is nine. What are the two numbers?

17. _____

18. How many liters of a 25% solution need to be mixed with a 40% solution in order to get 15 liters of a 35% solution?

18. _____

19. The perimeter of a rectangle is 92, if the width of the rectangle is four more than half the length, what are the dimensions of the rectangle.

19. _____

20. A woman invested half as much money at 4% annual interest as she did at 10% annual interest. If her total interest is $432, how much does she have invested in each account?

20. _____

Practice Test B

1. Is $(1,4)$ a solution of $\begin{cases} y = 3x+1 \\ 4x+y = -8 \end{cases}$

    a. Yes   b. No

1. _____

2. Is $(-3,-5)$ a solution of $\begin{cases} 3x - 2y = 1 \\ -2x = -2y - 4 \end{cases}$

    a. Yes   b. No

2.. _____

3. Solve the system by graphing. $\begin{cases} x + 6y = -2 \\ -x - 4y = 6 \end{cases}$

    a. (-10, -2)   b. (10, -2)

    c. No Solution   d. Infinite Solution

3. _____

4. Solve the system by graphing. $\begin{cases} y = \dfrac{1}{2}x - 4 \\ y = \dfrac{3}{4}x + 5 \end{cases}$

    a. (-8, -5)   b. (8, 5)

    c. (-8, 5)   d. (8, -5)

4. _____

5. Solve the system by the substitution method. $\begin{cases} x + y = 7 \\ y - x = -2 \end{cases}$

    a. $\left(\dfrac{5}{2}, \dfrac{9}{2} \right)$ b. $\left(\dfrac{9}{2}, \dfrac{5}{2} \right)$

    c. No Solution   d. Infinite Solution

5. _____

6. Solve the system by the substitution method. $\begin{cases} 7x - 4y = 8 \\ 3x - 9y = 18 \end{cases}$

    a. (0, -2)   b. (0, 2)

    c. No Solution   d. Infinite Solution

6. _____

7. Solve the system by the elimination method. $\begin{cases} x = 3 - y \\ y = -x + 5 \end{cases}$ 7. _____

    a. (- 4, 1)   b. (4, 1)

    c. No Solution   d. Infinite Solution

8. Solve the system by the substitution method. $\begin{cases} \dfrac{2}{3}x + \dfrac{5}{6}x = -1 \\ x = 4y \end{cases}$ 8. _____

    a. $\left(-\frac{8}{7}, -\frac{2}{7}\right)$ b. $\left(\frac{8}{7}, -\frac{2}{7}\right)$

    c. No Solution   d. Infinite Solution

9. Solve the system $\begin{cases} -3x + 4y = -10 \\ 6x - 8y = -12 \end{cases}$ 9. _____

    a. $\left(-\frac{8}{3}, -\frac{1}{2}\right)$ b. $\left(\frac{2}{3}, -2\right)$

    c. No Solution   d. Infinite Solution

10. Solve the system $\begin{cases} y = 2x - 6 \\ x = -\dfrac{1}{4}y - 3 \end{cases}$ 10. _____

    a. (-8, -1)   b. (-1, -8)

    c. No Solution   d. Infinite Solution

11. Solve the system $\begin{cases} 14x + 18y = 10 \\ 20x + 32y = -24 \end{cases}$ 11. _____

    a. $\left(5, -\frac{15}{4}\right)$ b. $\left(0, \frac{5}{8}\right)$

    c. No Solution   d. Infinite Solution

12. Graph the inequality. $2x - 5y > -10$

12. _____

a.
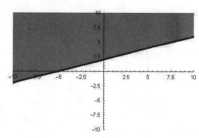

b.
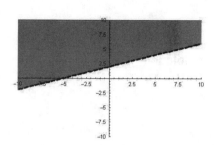

c.
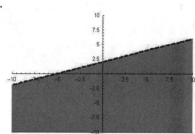

d.
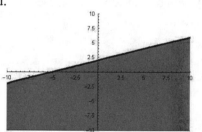

13. Graph the inequality. $x \le -2y - 6$

13. _____

a.
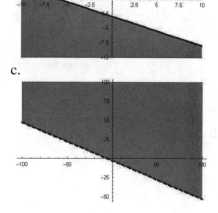

b.

c.

d.
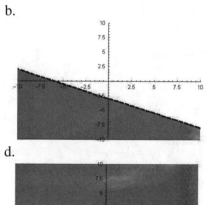

14. Graph the inequality. $0.25x - 0.60y \geq -0.15$ 14. _____

a.

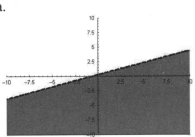

b.

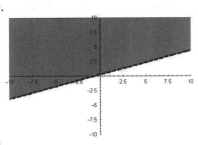

c.

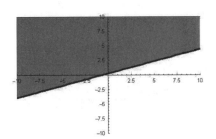

d.

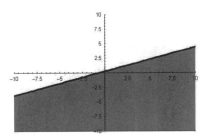

15. Graph the solutions of the system of linear inequality. $\begin{cases} y > \dfrac{3}{4}x - 2 \\ 3x - 4y \leq 12 \end{cases}$

a.

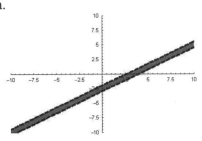

b.

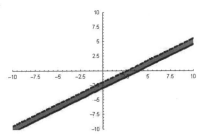

c.

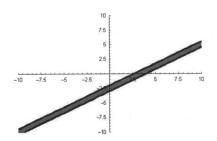

d.

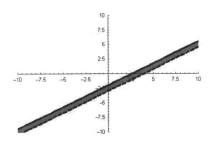

15. _____

16. Graph the solutions of the system of linear inequality. $\begin{cases} x+3y>5 \\ -x-5y<-8 \end{cases}$

a.

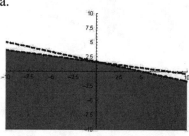

b.

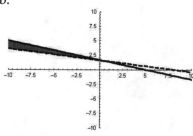

c.

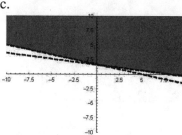

d.

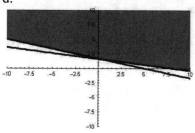

16. _____

17. One number is 4 more than three times another. The sum of the two numbers is 32. Find the numbers.

    a. 12 and 20   b. 11 and 21

17. _____

    c. 7 and 25   d. 18 and 14

18. David has $1.95 in dimes and nickels. He has a total of 22 coins. How many of each kind does he have?

    a. 17 nickels and 5 dimes   b. 5 nickels and 17 dimes

18. _____

    c. 10 nickels and 12 dimes   d. 12 nickels and 10 dimes

19. A 25-inch chain needs to be cut into two pieces so that one piece is 4 inches longer than twice the other, how long is each piece?

       a. 7 inches and 18 inches

                 b. 15 inches and 10 inches

19. _____

       c. 5 inches and 20 inches

                 d. 8 inches and 17 inches

20. Two angles are supplementary, and the measure of one is 30° less than three times the other. Find the measure of each angle.

       a. 127.5° and 52.5°

                 b. 128° and 52°

20. _____

       c. 127° and 53°

                 d. 130° and 50°

Chapter 5 Exponents and Polynomials
5.1 Exponents

Learning Objectives
1. Evaluate exponential expressions.
2. Use the product rule for exponents.
3. Use the power rule for exponents.
4. Use the power rules for products and quotients.
5. Use the quotient rule for exponents and define a number raised to the zero power.
6. Decide which rule(s) to use to simplify each expression.

Vocabulary.
Use the choices to complete each statement.

Add	Base	Divide	Exponents
Exponent	False	Multiply	Subtract
True	0	1	

1. In the expression 6^9, six is the _____ and nine is the _____.

2. According to the quotient rule, to simplify $x^{10} \div x^3$, we will _____ the exponents.

3. If given the expression $\left(x^5\right)^3$, to simplify we would _____ the exponents.

4. True or False. $(-3)^0 = -1$. _____

5. If there is no exponent, i.e. x, the exponent is understood to be _____.

Objective 1
Evaluate each expression.

6. $\left(\dfrac{3}{4}\right)^4$

6. _____

7. $(0.03)^3$

7. _____

8. -16^2

8. _____

Objective 2
Use the product rule to simplify. Write the results using exponents.

9. $\left(2^3 x^5\right)\left(2^8 x^{13}\right)$

9. _____

10. $m^{27}n^8\left(m^{10}n^{21}\right)$

10. _____

11. $\left(x^9y\right)\left(x^{10}y^5\right)$

11. _____

Objective 3
Use the power rule to simplify. Write the results using exponents.

12. $\left(5^3\right)^8$

12. _____

13. $\left[(-7)^4\right]^{12}$

13. _____

14. $\left(x^4\right)^{18}$

14. _____

Objective 4
Use the power rule and power of the product rule to simplify each expression.

15. $\left(a^4b^5\right)^6$

15. _____

16. $\left[(-4)^3 x^4\right]^9$

16. _____

17. $\left(0.2y^{16}\right)^3$

17. _____

Objective 5
Use the power rule and the power of a quotient to simplify each expression.

18. $\left(\dfrac{x}{6}\right)^4$

18. _____

 19. $\left(\dfrac{-2xz}{y^5}\right)^2$

19. _____

20. $\left(\dfrac{a^2b^4}{3c^5}\right)^5$

20. _____

21. $(-8)^0$

21. _____

22. $-14x^0$

22. _____

23. $\dfrac{m^0}{9}$

23. _____

Objective 6
Simplify each expression.

 24. $\dfrac{9a^4b^7}{27ab^2}$

24. _____

25. $\dfrac{\left(4m^3n^5\right)^3}{\left(2m^4n^6\right)^2}$

25. _____

Concept Extensions

26. Prove the fact that $7^0 = 1$. Hint: the quotient rule.

26. _____

27. The volume of a cube is given but he formula $V = x^3$, where x is the length of the side. How would the volume of a cube change if the side was quadrupled?

27. _____

Section 5.2 Adding and Subtracting Polynomials

Learning Objectives
1. Define polynomial, monomial, binomial, trinomial, and degree.
2. Find the value of a polynomial given replacement values for the variables.
3. Simplify a polynomial by combining like terms.
4. Add and subtract polynomials.

Vocabulary
Use the choices to complete each statement.

Binomial	**Coefficient**	**Constant**
False	**Monomial**	**True**

1. A two term polynomial is referred to as a(n) _____.
2. A numerical term in a polynomial without any variables is called a(n) _____.
3. True or false. The degree of the term is the greatest degree of the polynomial.
4. A single term polynomial is called a(n) _____.
5. The numerical _____ is the numerical factor of the term.

Objectives 1
Find the degree of each of the following polynomials and determine whether it is a monomial, binomial, trinomial, or none of these.

6. $5x^2 + 7$

6. _____

 7. $12x^4y - x^2y^2 - 12x^2y^4$

7. _____

8. $3m^2n + 4mn^2 - 7m^2n^2 - mn$

8. _____

9. $\frac{1}{2}x^2y - \frac{1}{4}y^3 - \frac{1}{8}xy^4 + \frac{1}{3}x^5y^3$

9. _____

10. $3a^2c^3 - 2c$

10. _____

                                           Martin-Gay *Beginning Algebra, Fifth Edition*

Objective 2
Find the value of the polynomial when x = -1.

11. $x^2 + 2x + 1$

11. _____

12. $3x^4 - 4x^3 + 2x^2 - 5$

12. _____

13. $6x^2 - 7x + 9x^3$

13. _____

Objective 3
Simplify each of the following by combining like terms.

14. $3x^4 - 5x^3 + 7x - 6x^3 - x^4$

14. _____

15. $5ab - 6a + 10ba - 8b$

15. _____

Objective 4
Perform the indicated operations.

16. $(4x - 3) + (8x + 7)$

16. _____

17. $(7x - 5) + (-17x - 8)$

17. _____

18. $(2x^2 + 5) - (3x^2 - 9)$

18. _____

19. $\left(3x^3 + 4x - 1\right) + \left(5x^3 - 4x^2\right) - \left(7x^3 + 3x^2 - 4\right)$

19. _____

20.

20. _____

$$5x^3 - 4x^2 + 6x - 2$$
$$-\left(3x^3 - 2x^2 - x - 4\right)$$

21. Subtract $\left(6m^2 - 7m + 1\right)$ from the sum of $\left(4m^2 - 3m\right)$ and $\left(8m - 9\right)$.

21. _____

22. $\left[\left(3x^2 - 5x + 4\right) - \left(5x^2 - 7x - 2\right)\right] + \left(6x^2 - 12x - 8\right)$

22. _____

Concept Extensions

23. An object is thrown from a cliff. The height of the object is given by the formula $-16t^2 - 32t + 496$; where t represents time. How high is the object after 4 seconds?

23. _____

Objective 5.3 Multiplying Polynomials

Learning Objectives

1. Use the distributive property to multiply polynomials.
2. Multiply polynomials vertically.

Objective 1
Multiply.

1. $2x^2\left(4x^3 - 3x\right)$

1. _____

2. $-\dfrac{1}{2}y\left(4y^4 - 8y^3 - 16y + 10\right)$

2. _____

3. $-y\left(4x^3 - 7x^2y + xy^2 + 3y^3\right)$

3. _____

4. $(x-7)(x+4)$

4. _____

5. $\left(\dfrac{1}{2}x - \dfrac{1}{3}\right)\left(\dfrac{1}{5}x - \dfrac{1}{4}\right)$

5. _____

6. $(m+3)(m-2)$

6. _____

7. $(a+7)(b-10)$

7. _____

8. $(x-7)^2$

8. _____

 9. $(3x^2+1)^2$

9. _____

10. $(x+2)(x^2-3x-4)$

10. _____

11. $(x-2)^3$

11. _____

12. $(m^2-3m+1)(m^2+4m-2)$

12. _____

Objective 52
Multiply vertically.

13. $(a-4)(a+1)$

13. _____

14. $(3m+n)(4m-3n)$

14. _____

 15. $(5x+1)(2x^2+4x-1)$

15. _____

16. $\left(y^2 - 3y + 4\right)\left(y^2 + 2y - 1\right)$

16. _____

Concept Extensions.

17. Write a polynomial that describes the area of the shaded area.

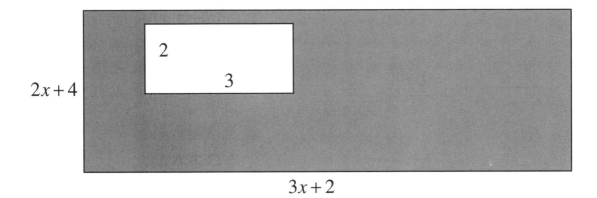

17. _____

Section 5.4 Special Products.

Learning Objectives
  1. Multiply two binomials using the FOIL method.
  2. Multiply the sum and difference of two terms.

Objective 5.4.1
Multiply using the FOIL method.

1. $(x-7)(x+6)$

1. _____

2. $(m+3)(m+9)$

2. _____

3. $(3y-4x)(11y-12x)$

3. _____

4. $(14a-9)(-a+7)$

4. _____

5. $(-2m-p)(-8m+p)$

5. _____

6. $(3b+7)(2b-15)$

6. _____

Objective 2
Multiply.

7. $\left(x-\dfrac{1}{4}\right)^2$

7. _____

8. $(5x+9)^2$

8. _____

Martin-Gay *Beginning Algebra, Fifth Edition*

9. $(7s - 3t)^2$

9. _____

10. $\left(\dfrac{1}{3}x - \dfrac{1}{2}y\right)^2$

10. _____

Objective 3
Multiply.

11. $(x + 5)(x - 5)$

11. _____

12. $\left(\dfrac{1}{2}x - 1\right)\left(\dfrac{1}{2}x + 1\right)$

12. _____

13. $(9x + y)(9x - y)$

13. _____

14. $(7x - 1.5)(7x + 1.5)$

14. _____

15. $(9x - 13)(9x + 13)$

15. _____

Concept Extensions.

16. $\big[(2x - 1) + 3y\big]\big[(2x - 1) - 3y\big]$

16. _____

17. $(5x - 2)^4$

17. _____

Section 5.5 Negative Exponents and Scientific Notation

Learning Objectives
1. Simplify expressions containing negative exponents.
2. Use all the rules and definitions for exponents to simplify exponential expressions.
3. Write numbers in scientific notation.
4. Convert numbers from scientific notation to standard form.

Vocabulary
Use the choices to complete each statement.

Standard notation   **Scientific notation**

$\dfrac{1}{x^7}$ x^7

1. 3.2×10^{-4} is written in _____

2. The expression x^{-7} equals _____.

3. The expression $\dfrac{1}{x^{-7}}$ equals _____.

4. 0.000032 is written in _____.

Objective 1
Simplify each expression. Write each result using positive exponents.

5. 8^{-3}

5. _____

6. $\dfrac{1}{7^{-5}}$

6. _____

7. $-4x^{-6}$

7. _____

8. $\dfrac{m^{-3}}{n^{-4}}$

8. _____

9. $3^{-2} - 3^0$

9. _____

Objective 2
Simplify each expression. Write each result using positive exponents.

 10. $\dfrac{r}{r^{-3}r^{-2}}$

10. _____

11. $\dfrac{\left(x^3\right)^{-4}}{x^{-8}}$

11. _____

12. $\dfrac{\left(2x^4y^5\right)^{-2}}{\left(4x^7y^3\right)^{-3}}$

12. _____

13. $\left(-4a^4b^3\right)\left(-2a^{-2}b^4\right)^2$

13. _____

Objective 5.5.3
Write each number in scientific notation.

14.   897,000,000

14. _____

15.   0.00000167

15. _____

16.   17,000,000,000

16. _____

Objective 4
Write each number in standard notation.

17. 3.478×10^{-7}

17. _____

Martin-Gay *Beginning Algebra, Fifth Edition*   169

18. 8.7325×10^8

18. _____

Concept Extensions.

Evaluate each expression using exponential rules. Write each result in standard notation.

 19. $\dfrac{1.4 \times 10^{-2}}{7 \times 10^{-8}}$

19. _____

20. $\left(3.2 \times 10^{-3}\right)\left(6.4 \times 10^4\right)$

20. _____

Section 5.6 Dividing Polynomials.

Learning Objectives.
  1. Divide a polynomial by a monomial.
  2. Use long division to divide a polynomial by another polynomial.

Objective 1
Perform each division.

1. $\left(6x^3 - 5x^2 + 4x\right) \div (-2x)$

1. _____

2. Divide $\left(24a^4b^3 + 32a^4b^4 - 16a^2b^2\right)$ by $4a^2b$

2. _____

 3. $\dfrac{-9x^5 + 3x^4 - 12}{3x^3}$

3. _____

4. $\dfrac{14m^2 - 8m - 6}{-4m}$

4. _____

Objective 2
Find each quotient using long division.

 5. $\dfrac{x^2 + 4x + 3}{x + 3}$

5. _____

6. $\dfrac{12x^2 - 11x - 3}{4x - 1}$

6. _____

7. $\left(8m^3 + 24m^2 + 2m - 14\right) \div \left(2m + 3\right)$

7. _____

8. $\dfrac{x^3 - 1}{x - 1}$

8. _____

9. $\dfrac{2b^3 + 9b^2 + 6b - 4}{b + 4}$

9. _____

10. $\dfrac{7 - 11x^2}{x + 1}$

10. _____

11. $\dfrac{m^5 - m^2}{m^4 - m}$

11. _____

Concept Extensions
Divide using long division.

12. $\dfrac{x^4 - x^3 - 4x^2 + 19x - 15}{x^2 - 3x + 5}$

12. _____

Chapter 5 Vocabulary

Exponent Rules: Power Rule: Product Rule: Power of a Product Rule: Power of a Quotient Rule: Quotient Rule: Zero Exponent:	$\left(x^m\right)^n = x^{m \cdot n}$ $x^m \cdot x^n = x^{m+n}$ $\left(xy\right)^n = x^n y^n$ $\left(\dfrac{x}{y}\right)^n = \dfrac{x^n}{y^n}$ $x^m \div x^n = x^{m-n}$ $x^0 = 1$	$\left(2^3\right)^4 = 2^{12}$ $2^3 \cdot 2^4 = 2^7$ $\left(2y\right)^4 = 2^4 y^4$ $\left(\dfrac{2}{y}\right)^4 = \dfrac{2^4}{y^4}$ $2^5 \div 2^2 = 2^3$ $2^0 = 1$
Polynomial	Finite sum of terms in the form ax^2	$8x^3 + 3x^2 - 2x - 9$
Monomial	A one term polynomial	$6c$
Binomial	A two term polynomial	$6x - 4$
Trinomial	A three term polynomial	$3x^2 - 2x - 9$
Degree of term	Sum of the exponents on the variables	$3x^2 y^4$ $2 + 4 = 6$
Degree of polynomial	The greatest degree of terms of the polynomial	$3x^2 - 2x - 9$ Degree of terms were 2, 1, and 0. Therefore the degree of polynomial is 2.
FOIL	Method to multiply binomials. First, Outer Inner, Last	$(x-1)(x+2)$ F: x^2 O: +2x I: - x L = -2

Squaring a binomial	$(a \pm b)^2 = a^2 \pm 2ab + b^2$	$(x-1)^2 =$ $x^2 - 2(x) + (1)^2 =$ $x^2 - 2x + 1$
Multiplying a Sum and Difference	$(a+b)(a-b) = a^2 - b^2$	$(x-3)(x+3) =$ $x^2 - (3)^2 =$ $x^2 - 9$
Scientific Notation	Method to write extremely large numbers or small decimals in a simplified manner.	$1,250,000,000 = 1.25 \times 10^9$ $0.0000000657 = 6.57 \times 10^{-8}$

Practice Test A

Evaluate each expression.

1. $\left(\dfrac{1}{8}\right)^3$

1. _____

2. -2^6

2. _____

Simplify each exponential expression. Write the results using only positive exponents.

3. $\dfrac{\left(r^{-2}s^3\right)^{-5}}{2r^{-5}s^4}$

3. _____

4. $\left(-3m^4n\right)^3\left(4m^3n^4\right)$

4. _____

Express each number in scientific notation.
5.   137,200,000,000

5. _____

Write each number in standard form.
6. 4.284×10^{-8}

6. _____

Simplify.

7. $\dfrac{1.5\times10^{-8}}{3\times10^{-9}}$

7. _____

Simplify by combining like terms.
8. $7x^2y+13xy^2-6x^2y+2y^2-8xy^2$

8. _____

Perform the indicated operation.
9. $\left(3a^2-4a-2\right)+\left(6a^2+10a\right)$

9. _____

10. Subtract $\left(7x^2 - 8x\right)$ from $\left(9 - 3x^2\right)$

10. _____

11. Subtract $\left(c^2 - 2c + 10\right)$ from the sum of $\left(3c^2 - 8c + 9\right)$ and $\left(-4c^2 + 8c - 5\right)$.

11. _____

Multiply.

12. $\left(a + b\right)\left(2a - 3b\right)$

12. _____

13. $-\dfrac{1}{4}x\left(16x^3 - 20x^2 + 8x - 32\right)$

13. _____

14. $\left(2m + n\right)\left(2m^2 - 8m + n\right)$

14. _____

15. $\left(x - 13\right)\left(x + 13\right)$

15. _____

16. $\left(y - 7\right)^2$

16. _____

17. $\left(7x - \dfrac{1}{2}\right)\left(7x + \dfrac{1}{2}\right)$

17. _____

18. $\left(c + 4\right)^3$

18. _____

19. Find the area of a rectangle if the width is five less than twice the length. Given that the length is x, state your answer in terms of x.

19. _____

20. The height of a falling object is given by the equation $-16x^2 + 324$, where x is time in seconds. How far does the object fall in 6 seconds?

20. _____

21. $\dfrac{4x^3 - 8x^2 - 6x}{2x}$

21. _____

22. $\dfrac{6m^2n - 8mn^2 - 16m}{8mn}$

22. _____

23. $\dfrac{2x^2 + x - 21}{x - 3}$

23. _____

24. $\dfrac{x^3 - 125}{x - 5}$

24. _____

25. $\left(6x^3 + x^2 - 10x + 8\right) \div \left(2x - 1\right)$

25. _____

Practice Test A

Evaluate each expression.
1. -0.04^3

      a. 0.000064   b. -0.12

      c. -0.000064   d. 0.12

1. _____

2. $(-1)^6$

      a. 6   b. -6

      c. -1   d. 1

2. _____

Simplify each exponential expression. Write the results using only positive exponents.

3. $\dfrac{\left(x^5 y^{-4}\right)^{-3}}{5y^6 x^{-4}}$

      a. $\dfrac{x^{11} y^6}{5}$ b. $\dfrac{x^{19}}{5y^{18}}$

3. _____

      c. $\dfrac{x^{11}}{5y^{10}}$ d. $\dfrac{y^6}{5x^{11}}$

4. $\left(-7s^9 r\right)\left(-5s^2 r^3\right)^2$

      a. $35s^{11} r^4$ b. $70s^{13} r^6$

4. _____

      c. $35s^{13} r^7$ d. $-175s^{13} r^7$

Express each number in scientific notation.
5.   65,790,000,000

      a. 6.579×10^{-10} b. 6.579×10^{10}

5. _____

      c. 6579×10^{-7} d. 6579×10^7

Write each number in standard form.

6. 9.32×10^{-4}

    a. 93,200

    b. 9,320,000

    c. 0.000932

    d. 0.0000932

6. _____

Simplify.

7. $\left(1.2 \times 10^{12}\right)\left(6 \times 10^{-8}\right)$

    a. 7.2×10^{4}

    b. 5.0×10^{4}

    c. 7.2×10^{-4}

    d. 7.2×10^{20}

7. _____

Simplify by combining like terms.

8. $5x^2 - 3y^2 + 6x - 8y + 14x^2 - 16y$

    a. $19x^2 + 6x - 24y - 3y^2$

    b. $-9x^2 + 6x - 8y - 3y^2$

    c. $25x - 27y$

    d. $-4x^2 y^2$

8. _____

Perform the indicated operation.

9. $\left(7x - 4y\right) + \left(-13x - 8y\right)$

    a. $20x - 12y$

    b. $20x + 4y$

    c. $-6x - 12y$

    d. $-6x + 4y$

9. _____

10. Subtract $\left(12m - 8m^2\right)$ from $\left(11m^2 + 12m\right)$

    a. $3m^2$

    b. $3m^2 + 24m$

    c. $19m^2 + 24m$

    d. $19m^2$

10. _____

11. Subtract $\left(2x^2 - 5x + 13\right)$ from the sum of $\left(8x^2 - 11x - 7\right)$ and $\left(-13x^2 - 4x + 17\right)$.

    a. $9x^2 - 15x - 10$ b. $-7x^2 - 10x - 3$

                                                                       11. _____

    c. $-5x^2 - 15x + 10$ d. $-7x^2 - 20x + 23$

Multiply.

12. $\left(x - 2y\right)\left(4x + 3y\right)$

        a. $4x^2 - 5xy - 6y^2$ b. $4x^2 + 5xy + 6y^2$

                                                                            12. _____

        c. $4x^2 + 11xy - 6y^2$ d. $4x^2 - 11xy - 6y^2$

13. $-5x\left(2x^3 + 10x^2 - 11x + 16\right)$

        a. $-3x^4 + 5x^3 - 16x^2 - 11x$ b. $-10x^3 - 50x^2 + 55x - 90$

                                                                            13. _____

        c. $-10x^4 - 50x^3 + 55x^2 - 90x$ d. $-3x^3 + 5x^2 - 16x - 11$

14. $\left(8x - 1\right)\left(2x^2 + 3x + 4\right)$

        a. $16x^3 + 26x^2 + 35x + 4$ b. $16x^3 + 26x^2 - 29x - 4$

                                                                            14. _____

        c. $16x^3 - 22x^2 - 35x - 4$ d. $16x^3 + 22x^2 + 29x - 4$

15. $\left(2x - 7\right)\left(2x + 7\right)$

        a. $4x^2 - 49$ b. $4x^2 - 28x - 49$

                                                                         15. _____

        c. $4x^2 + 49$ d. $4x^2 + 28x - 49$

16. $(w+3)^2$

    a. $w^2 + 9$ b. $w^2 + 6$

                                                16. _____

    c. $w^2 + 6w + 9$ d. $w^2 + 6w + 6$

17. $(x-0.25)(x+0.25)$

    a. $x^2 + 0.50x + 0.0625$ b. $x^2 - 0.0625$

                                                  17. _____

    c. $x^2 + 0.0625$ d. $x^2 - 0.50x - 0.0625$

18. $(x-2)^3$

    a. $x^3 - 6x^2 + 12x - 8$ b. $x^3 + 6x^2 + 12x + 8$

                                                  18. _____

    c. $x^3 - 8x - 8$ d. $x^3 - 16x^2 - 2x + 16$

19. Find the area of a triangle if the base is four less than double the height. Given that the height is x, state your answer in terms of x.

    a. $x^2 - 2$ b. $2x^2 - 4x$

    c. $2x^2 - 4$ d. $x^2 - 2x$

                                                   19. _____

20. The height of a falling object is given by the equation $-16x^2 + 1648$, where x is time in seconds. How far does the object fall in 8 seconds?

      a. -376 feet   b. 520 feet

      c. 1556 feet   d. 624 feet

20. _____

21. $\dfrac{16m^5 - 24m^2 - 8}{4m^2}$

      a. $4m^7 - 6m^4 - 2$ b. $4m^3 - 6m - 2$

21. _____

      c. $4m^3 - 6 - \dfrac{2}{m^2}$ d. $4m^3 + 6 + \dfrac{2}{m^2}$

22. $\dfrac{5x^4 - 7x^2 - 15x}{5x^2}$

      a. $x^2 - \dfrac{7}{5}x - 3$ b. $x^2 + \dfrac{7}{5} + \dfrac{3}{x}$

22. _____

      c. $x^2 - \dfrac{7}{5} - \dfrac{3}{x}$ d. $x^4 - \dfrac{7}{5}x^2 - 3x$

23. $\dfrac{8x^2 - 2x - 3}{2x + 1}$

     a. $4x + 3$ b. $4x - 3$

                                                          23. _____

     c. $4x - 1 - \dfrac{3}{2x}$ d. $4x + 3 + \dfrac{2}{2x + 1}$

24. $\dfrac{8x^3 - 1}{2x - 1}$

     a. $4x^2 + 1$ b. $4x^2 - 2x - 1$

                                                           24. _____

     c. $4x^2 - 2x + 1$ d. $4x^2 + 2x + 1$

25. $\left(4x^2 - 29x + 40\right) \div \left(x - 6\right)$

     a. $4x - 5 + \dfrac{10}{x - 6}$ b. $4x - 5$

                                                           25. _____

     c. $4x + \dfrac{-5x + 40}{x - 6}$ d. $4x + 5 - \dfrac{10}{x - 6}$

Chapter 6 Factoring Polynomials
Section 6.1 The Greatest Common Factor and Factoring by Grouping

Learning Objectives
1. Find the greatest common factor of a list of integers.
2. Find the greatest common factor of a list of terms.
3. Factor out the greatest common factor from a polynomial.
4. Factor a polynomial by grouping.

Vocabulary
Use the choices to complete each statement.

Factoring **Factors** **False**
Greatest common factor **True**

1. _____ is what is multiplied together to result in a product.

2. True or False. A number that does not have any factors other than one and itself is a composite

    number. _____.

3. The _____ is the largest factor that can divide into all the terms evenly.

4. _____ is the mathematical process of rewriting a polynomial as a product.

Objective 1

Find the GCF for each list.

5. 18, 27, 36

5. _____

6. 64, 82, 108

6. _____

7. 45, 39, 19

7. _____

Objective 2

Find the GCF for each list

8. m^8, m^6, m^4

8. _____

9. $12y^4$ and $20y^3$

9. _____

10. $-40x^2y^3$, $20x^4y^2$, $16x^3y^3$

10. _____

11. $15a^7b^5$, $25a^4b^3$, $27a^8b^9$

11. _____

Objective 3

Factor out the GCF from each polynomial.

12. $14x - 7$

12. _____

13. $9x^2 - 18x + 6$

13. _____

14. $m^4n^3 - m^3n^3 + m^5n^4 - m^6n^7$

14. _____

 15. $y(x^2 + 2) + 3(x^2 + 2)$

15. _____

Objective 4

Factor the given polynomials by grouping.

16. $5xy - 15x - 6y + 18$

16. _____

17. $ab + 5b^2 - 4a - 20b$

17. _____

18. $21xy - 56x + 6y - 16$

18. _____

Concept Extension

Factor.

19. $168ab + 96a - 189b - 108$

19. _____

20. $24y^{2n} - 60y^n - 20y^n + 50$

20. _____

Section 6.2 Factoring Trinomials of the Form $x^2 + bx + c$

Learning Objectives
1. Factor trinomial of the form $x^2 + bx + c$
2. Factor out the greatest common factor and then factor trinomial of the form $x^2 + bx + c$

Objective 1

Factor each trinomial.

1. $x^2 - 3x - 18$

1. _____

2. $x^2 - 11x + 28$

2. _____

3. $x^2 + 12x + 32$

3. _____

4. $x^2 - 13x - 36$

4. _____

5. $x^2 - 2x - 3$

5. _____

6. $x^2 - 5x + 6$

6. _____

7. $x^2 - xy - 20y^2$

7. _____

Objective 2

Factor each trinomial completely.

8. $3x^2 + 9x - 30$

8. _____

9. $x^2 - 7x + 44$

9. _____

10. $5x^3y - 25x^2y^2 - 120xy^3$

10. _____

11. $16x^2y - 16xy - 180y$

11. _____

12. $-4x^4 + 8x^2 + 60$

12. _____

13. $\frac{1}{2}x^2 - 2x - 16$

13. _____

Concept Extension

Factor completely.

14. $t^2(s-1) - 3t(s-1) - 28(s-1)$

14. _____

15. Find a positive value for b so that this trinomial is factorable. $x^2 + bx - 44$

15. _____

16. Find a positive value for c so that this trinomial is factorable. $x^2 - 17x + c$

16. _____

Martin-Gay *Beginning Algebra, Fifth Edition*

Section 6.3 Factoring Trinomials of the Form $ax^2 + bx + c$ **and Perfect Square Trinomials**

Learning Objectives
  1. Factor trinomial of the form $ax^2 + bx + c$ where $a \neq 1$
  2. Factor out the greatest common factor then factor trinomial of the form $ax^2 + bx + c$
  3. Factor a perfect square trinomial

Objective 1

Factor completely.

1. $6x^2 + x - 2$

1. _____

2. $10x^2 + 31x + 3$

2. _____

3. $5x^2 - 14x + 20$

3. _____

4. $3x^2 - 19x - 14$

4. _____

5. $27x^2 - 57x + 20$

5. _____

Objective 2

Factor completely.

6. $6x^2 - 27x - 15$

6. _____

7. $4x^3 - 9x^2 - 9x$

7. _____

8. $-16x^2 + 80x - 100$

8. _____

9. $56m^2 - 63 + 7m$

9. _____

10. $36x^2y + 21xy - 30y$

10. _____

Objective 3

Factor completely.

11. $16x^2 - 40x + 25$

11. _____

Martin-Gay *Beginning Algebra, Fifth Edition*

12. $9x^2 - 24xy + 16y^2$

12. _____

13. $16x^2 - 48x + 36$

13. _____

14. $64x^2 - 144x + 81$

14. _____

Concept Extension

Factor completely.

15. $36x^2 - 6x + \dfrac{1}{4}$

15. _____

16. $\dfrac{16}{49}x^2 - \dfrac{8}{21}xy + \dfrac{1}{9}y^2$

16. _____

17. $\dfrac{1}{4}x^2 - \dfrac{2}{3}x + \dfrac{1}{3}$

17. _____

Section 6.4 Factoring Trinomials of the Form $ax^2 + bx + c$ **by Grouping**

Learning Objective
1. Use the grouping method to factor trinomial of the form $ax^2 + bx + c$

Objective 1

Factor completely.

 1. $x^2 + 3x + 2x + 6$

1. _____

2. $30x^2 - 18x - 5x + 3$

2. _____

3. $2a^2 - 5a - 8a + 20$

3. _____

4. $4x^2 - 9x - 9$

4. _____

5. $12x^2 - 8x - 7$

5. _____

6. $9x^2 - 25x + 20$

6. _____

7. $36x^2 - 60x - 24$

7. _____

Martin-Gay *Beginning Algebra, Fifth Edition*

8. $20x^2y + 22xy - 12y$

8. _____

9. $12x^2 + 47xy + 40y^2$

9. _____

10. $21x^2 - 17x - 8$

10. _____

Concept Extension

Factor completely.

11. $\dfrac{1}{2}x^2 + \dfrac{11}{4}x - \dfrac{3}{2}$

11. _____

12. $x^2(x-7) - 16(x-7)$

12. _____

Section 6.5 Factoring Binomials

Learning Objectives
1. Factor the difference of two squares.
2. Factor the sum or difference of two cubes.

Vocabulary.
Use the choices to complete each statement.

Difference of cubes   **Difference of squares**
Perfect square trinomial   **Sum of cubes**

1. $25x^2 - 36$ is an example of _____.

2. $8x^3 + 64$ is an example of _____.

3. $(x+1)^2$ when FOILed out will result in a _____.

4. $125x^3 - 1$ is an example of _____.

Objective 1

Factor completely.

5. $x^2 - 49$

5. _____

6. $25x^2 - 144$

6. _____

 7. $121m^2 - 100n^2$

7. _____

8. $81x^2 - 64$

8. _____

9. $169x^2 + 36y^2$

9. _____

10. $x^8 - 1$

10. _____

11. $\dfrac{25}{36}t^2 - \dfrac{4}{9}$

11. _____

Objective 2

Factor completely.

 12. $x^3 + 125$

12. _____

13. $y^3 + 27$

13. _____

14. $b^3 - 216$

14. _____

15. $t^3 + 16$

15. _____

16. $xy^3 - 9xyz^2$

16. _____

17. $x^4 - y^6$

17. _____

18. $64t^3 - 729s^3$

18. _____

Concept Extension

Factor completely.

19. $(x-1)^3 - 27$

19. _____

20. $2.25t^2 - 1.69$

20. _____

Section 6.6 Solving Quadratic Equations by Factoring

Learning Objectives
1. Solve quadratic equation by factoring.
2. Solve equations with degree greater than two by factoring.
3. Find the x-intercept of a quadratic equation in two variables.

Objective 1

Solve the following equations.

1. $x^2 + x - 12 = 0$

1. _____

2. $2x^2 + 7x - 15 = 0$

2. _____

3. $6x^2 - 17x = -7$

3. _____

4. $2x^2 - 6x - 20 = 0$

4. _____

5. $x(x+14) = -49$

5. _____

6. $6x^2 + x - 15 = 0$

6. _____

Objective 3

Factor completely.

7. $5x^3 - 5x = 0$

7. _____

 8. $(2x+3)(4x-5) = 0$

8. _____

9. $x^3 - 5x^2 - 24x = 0$

9. _____

10. $y^3 + y^2 - 6y = 0$

10. _____

11. $-3x^3 + 6x^2 + 9x = 0$

11. _____

Objective 3

Find the x-intercept of the following quadratic equations.

12. $y = x^2 - 4x - 21$

12. _____

 13. $y = 2x^2 + 11x - 6$

13. _____

14. $y = 6x^2 - 5x - 4$

14. _____

15. $y = 3x^3 - 24x^2$

15. _____

Concept Extension

Solve.

16. An object is dropped from a height of 560 feet. The height of the object is given by the equation $h(t) = -16t^2 - 32t + 560$, at what time, t, will the object hit the ground?

16. _____

Section 6.7 Quadratic Equations and Problem Solving

Learning Objectives
1 . Solve problems that can be modeled by quadratic equations.

Objective 1

Solve.

 1. The perimeter of the triangle is 85 feet. Find the lengths of its sides.

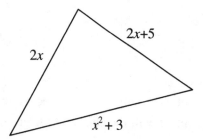

1. _____

2. The area of a square is 169 square units. Find the length of its sides.

2. _____

3. The product of two consecutive even integers is 168. Find the numbers.

3. _____

 4. The sum of a number and its square is 182. Find the number(s).

4. _____

5. The difference of two integers is 2, and their product is 143. Find the numbers.

5. _____.

6. An object is thrown upward from the top of an 80-foot building with an initial velocity of 64 feet per second. The height h of the object after t seconds is given by the quadratic equation $h = -16t^2 + 64t + 80$. When will the object hit the ground?

6. _____

7. The perimeter of a textbook is 28 inches. And the diagonal measures 10 inches. What are the length and width of the textbook?

7. _____

8. Melissa's age is twice Alice's. If the sum of the squares of their ages is 80, then what are their ages?

8. _____

9. The diameter of a circle is two less than the diameter of another circle. If the areas of the circles differ by 5π square meters, then what is the diameter of each circle?

9. _____

10. The length of the longer side of a right triangle is 2 feet less than the hypotenuse. The length on the shorter leg is 7 feet less than the length of the longer leg. Find the length of all three sides.

10. _____

Concept Extension

11. The sum of two numbers is 28. The sum of their squares is 400. Find the two numbers.

11. _____

12. Write down two numbers whose sum is 25. Square each number, and find the sum of their squares. Use this work to create your own word problem, like in number 11.

12. _____

Chapter 6 Vocabulary

Vocabulary Word	Definition	Example
Factoring	The process of writing an expression as a product.	$10 = 5 \cdot 2$ $x^2 + 3x + 2 = (x+2)(x+1)$
Greatest Common Factor	The biggest factor that goes into all the other terms.	2 is the GCF of 4, 18, and 22 x^2 is the GCF of x^2, x^4, x^7
Factor by grouping	Method to factor a polynomial, find the GCF between pairs of terms, then factor out a common binomial if present.	$2x^3 + x^2 + 6x + 3$ $x^2(2x+1) + 3(2x+1)$ $(2x+1)(x^3+3)$
Perfect Square Trinomial	$a^2 \pm 2ab + b^2$	$4x^2 - 12x + 9$
Difference of Squares	$a^2 - b^2$	$16x^2 - 25$
Sum of Cubes	$a^3 + b^3$	$8x^3 + 125$
Difference of Cubes	$a^3 - b^3$	$64x^3 - 1$
Quadratic Equation	$ax^2 + bx + c = 0$	$2x^2 - 3x + 2 = 0$
Zero Factor Theorem	If $a \cdot b = 0$, then $a = 0$ or $b = 0$.	$(x-2)(x+1) = 0$ $x - 2 = 0 \qquad x + 1 = 0$ $x = 2 \qquad\quad x = -1$

Practice Test A

Factor each polynomial completely.

1. $x^2 - 3x - 28$

1. _____

2. $6x^2 + 11x - 10$

2. _____

3. $8x^3 + 1$

3. _____

4. $3x^2 - 30x + 27$

4. _____

5. $x^4 - 16$

5. _____

6. $-4x^2 + 23x - 28$

6. _____

7. $3x^3 - 6x^2 - 5x + 10$

7. _____

Martin-Gay *Beginning Algebra, Fifth Edition*

8. $4x^2 - 169$

8. _____

9. $x^2 + 11x + 30$

9. _____

10. $16a^2 + 25$

10. _____

Solve.

11. $x^2 - 13x + 36 = 0$

11. _____

12. $2x^3 + 14x^2 + 12x = 0$

12. _____

13. $20x^2 + 26x - 6 = 0$

13. _____

14. $-15x^2 - 16x = 4$

14. _____

15. $25x^2 = -16 - 40x$

15. _____

16. $24x^2 - 20x - 4 = 0$

16. _____

Find the *x*-intercept of the given quadratic equation.

17. $y = 11x^2 + 21x - 2$

17. _____

18. $y = 100x^2 - 40x + 4$

18. _____

19. The difference of two numbers is 5. The sum of their squares is 125. Find the numbers.

19. _____

20. A compass is falling off a 324 foot cliff. If the height of the compass can be found using the equation, $h = -16t^2 + 324$, where *t* is time in seconds. At what time will the compass hit the ground?

20. _____

Practice Test B

Factor each polynomial completely.

1. $x^2 + 4x - 21$

    a. (x+7)(x-3)   b. (x-7)(x+3)

                                                1. _____

    c. (x − 7)(x − 3)   d. prime

2. $2x^2 + 9x - 18$

    a. (2x − 9)(x + 2)   b.(x + 6)(2x − 3)

                                                  2. _____

    c. (x + 9)(2x-2)   d. (x − 6)(2x + 3)

3. $125 - 64x^3$

    a. $(5 - 4x)(5+4x)^2$ b. $(5-4x)(25-20x+16x^2)$

                                                  3. _____

    c. $(5-4x)(25+20x+16x^2)$ d. prime

4. $x^3 + 3x^2 - 10x$

    a. x(x − 2)(x + 5)   b. $(x^2 - 2x)(x+5)$

                                                  4. _____

    c. (x − 2) (x^2+5x) d. prime

5. $y^2 - 11y + 30$

    a. (y −3)(y − 10)   b. (y − 5)(y − 6)

                                                  5. _____

    c. (y + 5)(y + 6)   d. prime

6.. $3x^2 - 12$

   a. $3(x^2-4)$ b. $3(x-2)^2$

6. _____

   c. prime   d. $3(x+2)(x-2)$

7. $12x^2 + 54x - 30$

   a. $(4x-3)(3x-10)$ b. $6(x+5)(2x-1)$

7. _____

   c. $6(2x-5)(x-1)$ d. prime

8. $x^2y + 3xy - 4y$

   a. $y(x-1)(x+4)$ b. $y(x-2)(x+2)$

8. _____

   c. $(xy-2)(x-2y)$ d. prime

9. $4x^2 - 4xy - 24y^2$

   a. $4(x-3y)(x+2y)$ b. $2(x-2y)(x+6y)$

9. _____

   c. $4(x-4y)(x-3y)$ d. prime

10. $15 - 5x - 3x^3 + 9x^2$

   a. $(x-3)(5-3x^2)$ b. $(x^2-3)(5-x)$

10. _____

   c. $3(x+5)(x-2)$ d. prime

                                Martin-Gay *Beginning Algebra, Fifth Edition*

Solve.

11. $2x^3 - 2x^2 - 40x = 0$

    a. 0, -10, 4   b. -5, -1, and 8

                                          11. _____

    c. 0, 5, -4 d. no solution

12. $4x^2 - 25 = 0$

    a. 5 and -5   b. 2 and - 2

                                          12. _____

    c. $-\frac{5}{2}$ and $\frac{5}{2}$ d. no solution

13. $3x^2 + 8x + 5 = 0$

    a. 1, and 5   b. $-\frac{5}{3}$ and -1

                                          13. _____

    c. -1 and -3   d. no solution

14. $3x^3 - 2x^2 - 8x = 0$

    a. 1, -3 and 8   b.1, 2, and 4

                                          14. _____

    c. no solution   d. 0, 2, and $-\frac{4}{3}$

15. $18y^3 + 3y^2 - 3y = 0$

    a. 6, -3, and 1   b. $0, -\frac{1}{2}$, and $\frac{1}{3}$

                                                      15. _____

    c. $1, -\frac{1}{4}$, and $\frac{2}{3}$ d. no solution

16. $x^2 + 7x + 12 = 0$

    a.   -12 and -1   b. - 4, - 3

                                                      16. _____

    c.   6 and 2   d. no solution

Find the x-intercept of the given quadratic equation.

17. $y = 10x^2 + 41x + 21$

    a.   -3 and -7   b. 7 and 3

                                                      17. _____

    c. $-\frac{3}{5}$ and $-\frac{7}{2}$ d. no x-intercept

18. $y = 6x^2 - 2x - 28$

    a.   2 and 7   b. -7 and 4

                                                     18. _____

    c.   no x-intercept   d. $\frac{7}{3}$ and -2

19. The area of a square is 144 square units. Find the length of a side

    a. 13   b. 12

    c. 14   d. 10

                                                    19. _____

20. Find two consecutive negative integers whose product is 240. Find the numbers

    a. -12 and - 20   b. – 24 and - 10

    c. -15 and -16   d. 12 and 20

                                                    20. _____

Chapter 7 Rational Expressions
Section 7.1 Simplifying Rational Expressions.

Learning Objectives
1. Find the value of a rational expression given a replacement number.
2. Identify values for which a rational expression is undefined.
3. Simplify or write rational expressions in lowest terms.

4. Write equivalent rational expressions of the form $-\dfrac{a}{b} = \dfrac{-a}{b} = \dfrac{a}{-b}$.

Objective 1

Find the value of the following expressions when $x = 4$, $y = -3$, and $z = -1$.

1. $\dfrac{y-3}{x+5}$

1. _____

2. $\dfrac{x^2 - z^3}{y^2}$

2. _____

3. $\dfrac{y^2 - 3y + 4}{x^2 + 3x - 4}$

3. _____

Objective 2

Find any numbers for which each rational expression is undefined.

 4. $\dfrac{x+3}{x+2}$

4. _____

5. $\dfrac{x-7}{4}$

5. _____

6. $\dfrac{x^2 - 4}{x^2 + 5x + 6}$

6. _____

7. $\dfrac{x^2 - 6x - 7}{4x^2 + 1}$

7. _____

Objective 3

Simplify each expressions.

8. $\dfrac{t + 1}{1 + t}$

8. _____

9. $\dfrac{11x + 121}{x^2 + 12x + 11}$

9. _____

10. $\dfrac{h - 2}{2 - h}$

10. _____

 11. $\dfrac{x^3 + 7x^2}{x^2 + 5x - 14}$

11. _____

12. $\dfrac{x^2 + x - 30}{2x^2 - 13x + 15}$

12. _____

Martin-Gay *Beginning Algebra, Fifth Edition*   215

Objective 4

List four equivalent forms for each rational expression.

 13. $-\dfrac{x+11}{x-4}$

13. _____

14. $-\dfrac{2y-5}{7-3y}$

14. _____

15. $-\dfrac{x-5}{9-x}$

15. _____

Concept Extension

16. Is the expressions $x-7$ and $\dfrac{x^2-9x-14}{x-2}$ equivalent for all real numbers? If not what values, would the two expressions not be equivalent.

15. _____

Section 7.2 Multiplying and Dividing Rational Expressions

Learning Objectives
1. Multiply rational expressions.
2. Divide rational expressions.
3. Multiply or divide rational expressions.

Objective 1

Find each product and simplify if possible.

1. $\dfrac{2w^2}{15x^3} \cdot \dfrac{25x}{6w}$

1. _____

2. $\dfrac{16t^2}{35} \cdot 70t^4$

2. _____

3. $\dfrac{x-2}{x+3} \cdot \dfrac{4}{x-2}$

3. _____

4. $\dfrac{4x^2-1}{2x^2+x-1} \cdot \dfrac{5x^2-5}{16x+8}$

4. _____

5. $\dfrac{z^2-1}{(z-1)^2} \cdot \dfrac{z-1}{z^2+2z+1}$

5. _____

6. $\dfrac{5x-20}{3x^2+x} \cdot \dfrac{3x^2+13x+4}{x^2-16}$

6. _____

Objective 2

Find each quotient and simplify.

7. $\dfrac{49x^2}{16y^3} \div \dfrac{28x^4}{32y^5}$

7. _____

8. $\dfrac{(x+3)(x-3)}{(x-7)(x-2)} \div \dfrac{x-7}{x-2}$

8. _____

9.

$$\dfrac{\dfrac{x+2}{x-3}}{\dfrac{x+2}{x-4}}$$

9. _____

 10. $\dfrac{x+2}{7-x} \div \dfrac{x^2-5x+6}{x^2-9x+14}$

10. _____

11. Find the quotient of $\dfrac{x^2-1}{x-4}$ and $\dfrac{x^2+2x+1}{x^2-7x+12}$.

11. _____

Objective 3

Multiply or divide as indicated.

12. $\dfrac{4x^2-9}{x^2+x-12} \cdot \dfrac{x^2+-x-6}{4x^2+12x+9}$

12. _____

13. $\dfrac{6x^2+5x-4}{2x^2+9x-5} \div \dfrac{3x^2-8x-16}{x^2+x-20}$

13. _____

14. Find the quotient of $\dfrac{x^2-7x-30}{x-5}$ and $\dfrac{x^2-6x-40}{x^2-10x+25}$.

14. _____

Concept Extension

Multiply or divide as indicated.

15. $\left(\dfrac{x^2+4x+3}{x^2+7x+10} \cdot \dfrac{x^2+4x-5}{x^2+2x-3} \right) \div \left(\dfrac{x^2+5x+4}{x^2+5x+6} \div \dfrac{x^2+2x-8}{x^2+2x-3} \right)$

15. _____

Section 7.3 Adding and Subtracting Rational Expressions with Common Denominators and Least Common Denominator

Learning Objectives
1. Add and subtract rational expressions with the same denominator.
2. Find the least common denominator of a list of rational expressions.
3. Write a rational expression as an equivalent expression whose denominator is given.

Objective 1

Add or subtract as indicated. Simplify if possible.

1. $\dfrac{x-3}{x+2} + \dfrac{2x-7}{x+2}$

1. _____

2. $\dfrac{-10}{x-6} + \dfrac{x+4}{x-6}$

2. _____

3. $\dfrac{x-4}{x+3} - \dfrac{7x}{x+3}$

3. _____

 4. $\dfrac{2x+3}{x^2-x-30} - \dfrac{x-2}{x^2-x-30}$

4. _____

5. $\dfrac{x^2-2x+3}{x+1} - \dfrac{x^2+3x-5}{x+1}$

5. _____

6. $\dfrac{2x^2+4x+3}{5x+15} + \dfrac{3x^2+x+2}{5x+15}$

6. _____

Objective 2

Find the LCD for each list of rational expressions.

7. $\dfrac{4}{x-3}, \dfrac{5}{x+1}$

7. _____

8. $\dfrac{x-2}{x^2+2x+1}, \dfrac{x+3}{x^2+5x+4}$

8. _____

9. $\dfrac{5x}{x^2-16}, \dfrac{7x+1}{x^2+8x+16}$

9. _____

 10. $\dfrac{1}{3x+3}, \dfrac{8}{2x^2+4x+2}$

10. _____

11. $\dfrac{3x+1}{x^3+1}, \dfrac{2x-7}{x^2-1}$

11. _____

Objective 3

Rewrite each rational expression as an equivalent rational expression with the given denominator.

12. $\dfrac{7}{a+1} = \dfrac{}{a^2-1}$

12. _____

13. $\dfrac{3x}{x(x-2)} = \dfrac{}{x^3 - 5x^2 + 6x}$

13. _____

 14. $\dfrac{9a+2}{5a+10} = \dfrac{}{5b(a+2)}$

14. _____

15. $\dfrac{x-1}{x^2+3x-10} = \dfrac{}{x^3+5x^2-4x-20}$

15. _____

Concept Extension

16. Write two rational expressions whose differences is $\dfrac{3x-9}{x^2-8}$.

16. _____

Name:

Date:

Instructor:

Section:

Section 7.4 Adding and Subtracting Rational Expressions with Unlike Denominators

Learning Objectives
1. Add and subtract rational expressions with unlike denominators.

Objective 1

Add or subtract as indicated.

1. $\dfrac{4}{x+1} + \dfrac{3x}{x^2-1}$

1. _____

2. $\dfrac{8a}{a-2} + \dfrac{a+1}{2a-4}$

2. _____

3. $\dfrac{y+2}{y+3} - 2$

3. _____

4. $\dfrac{3x+1}{x-1} - \dfrac{2x-3}{1-x}$

4. _____

5. $\dfrac{x-3}{x^2+6x+9} + \dfrac{x+9}{x^2+8x+15}$

5. _____

6. $\dfrac{x+1}{x^2+2x-48}-\dfrac{2x}{x^2-14x+48}$

6. _____

7. $\dfrac{2}{y^2+3y-4}-\dfrac{4}{y^2-y-12}$

7. _____

8. $\dfrac{4}{2x^2-7x-15}+\dfrac{x-2}{2x^2+17x+21}$

8. _____

9. $\dfrac{x+2}{x^2-7x-8}+\dfrac{x+1}{x^2-9x+8}$

9. _____

10. $\dfrac{y-8}{y^2-8y+12}-\dfrac{y+2}{y^2+4y-12}$

10. _____

11. $\dfrac{x+8}{x^2-5x-6}+\dfrac{x+1}{x^2-4x-5}$

11. _____

12. $\dfrac{x-3}{x^2-5x-14}-\dfrac{x+2}{x^2-10x-21}$

12. _____

Concept Extension

13. $\dfrac{3x+1}{x^2-8x+12}-\dfrac{2x}{x^2-4x-12}+\dfrac{4}{x^2-4}$

13. _____

Section 7.5 Solving Equations Containing Rational Expressions

Learning Objectives
1. Solve equations containing rational expressions.
2. Solve equations containing rational expressions for a specified variable.

Objective 1

Solve each equation and check each proposed solution.

1. $\dfrac{2}{x-1} = 5$

1. _____

2. $\dfrac{7}{x-2} + 2 = \dfrac{5}{x-2}$

2. _____

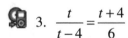 3. $\dfrac{t}{t-4} = \dfrac{t+4}{6}$

3. _____

4. $-1 + \dfrac{2y}{y+3} = \dfrac{-4}{y+4}$

4. _____

5. $\dfrac{x}{x-6} = \dfrac{6}{x-6} + 3$

5. _____

6. $\dfrac{2}{2n-3} - \dfrac{2}{10n^2 - 13n - 3} = \dfrac{n}{5n+1}$

6. _____

7. $\dfrac{2}{4x-3} + \dfrac{1}{3x+5} = \dfrac{7x+2}{12x^2 + 11x - 15}$

7. _____

8. $\dfrac{2x}{6x^2 + 7x - 3} + \dfrac{5}{2x^2 + 11x + 12} = \dfrac{x-3}{3x^2 + 11x - 4}$

8. _____

9. $\dfrac{y-2}{9y^2 - 1} - \dfrac{7}{6y-2} = -\dfrac{1}{3y-1}$

9. _____

10. $\dfrac{4r-4}{r^2 + 5r - 14} + \dfrac{2}{r+7} = \dfrac{1}{r-2}$

10. _____

Objective 2

Solve each equation for specified variable.

11. $\dfrac{1}{p} + \dfrac{1}{q} = \dfrac{1}{r}$ for p.

11. _____

12. $\dfrac{P - 2L}{W} = 2$ for L.

12. _____

 13. $T = \dfrac{2U}{B + E}$ for B.

13. _____

14. $L = \dfrac{S - 2WH}{2(W + H)}$ for H.

14. _____

Concept Extension

15. Why do you have to check your proposed solutions? Why are they not solutions of the equations?

15. _____

Section 7.6 Proportion and Problem Solving with Rational Equations

Learning Objectives
1. Solve proportions
2. Use proportions to solve problems
3. Solves problems about numbers
4. Solve problems about work.
5. Solve problems about distance.

Objective 1

Solve each proportion.

1. $\dfrac{x}{6} = \dfrac{11}{2}$

1. _____

2. $\dfrac{x+1}{2x+3} = \dfrac{2}{3}$

2. _____

3. $\dfrac{3}{x+1} = \dfrac{4}{x-1}$

3. _____

4. $\dfrac{8}{x-4} = \dfrac{2}{x+2}$

4. _____

Objective 2

Solve.

5. In a class of 50 students, 30 of them are women. In a class, of 100 students, how many are expected to be women?

5. _____

6. It has been advertised that 3 out of 5 dentists endorse the Great Whiting toothpaste. If there are 1525 dentists at this year's annual New Hampshire dentist convention, how many of them endorse the toothpaste?

6. _____

7. If there is 1000 meters in 1 kilometer, then how many meters are in 4.56 kilometers?

7. _____

Objective 3

Solve.

8. The quotient of 5 and the sum of a number and 3 is the same as 8 divided by 7 more than that number. Find the number.

8. _____

9. Twelve divided by the sum of x and two equals the quotient of 4 and the difference of x and 2. Find x.

9. _____

10. The ratio of 4 and the first of two consecutive even integers equals the ratio of 6 and the second integer. Find the two integers.

10. _____

Objective 4

Solve.

11. In 2 minutes, a conveyor belt moves 300 pounds of recyclable aluminum from the delivery truck to a storage area. A smaller belt moves the same quantity of cans the same distance in 6 minutes. If both belts are used, find how long it takes to move the cans to the storage area.

11. _____

12. Paul paints a room in 6 hours. Joe can paint a room in 3 hours. How long will it take them to paint a room together?

12. _____

13. It takes Joel 2.5 hours to mow the yard, when his wife helps they complete the task in 1.5 hours. how long will it take Joel's wife to mow the yard by herself.

13. _____

Objective 5

Solve.

14. A car travels 280 miles in the same time a motorcycle travels 240 miles. If the car's speed is 10 miles per hour more than the motorcycle's, find the speed of the car and the speed of the motorcycle.

14. _____

15. Michelle can walk 16 miles in the same time it takes Lucy to walk 8 miles. If Michelle walks 1 mph faster than Lucy, how fast does each person walk?

15. _____

16. A Cessna-172 will leave Chicago and fly 450 miles to it's destination. A car will leave Chicago at the same time going to the same destination. If the Cessna-172 maintains a speed of 3 times that of the car, the Cessna-172 will reach the destination 6 hours ahead of the car. Find the speed of the Cessna-172.

16. _____

Section 7.7 Variation and Problem Solving

```
Learning Objectives
   1. Solve problems involving direct variation.
   2. Solve problems involving inverse variation.
   3. Other types of direct and inverse variation.
   4. Variation and problem solving.
```

Objective 1

Write an equation to describe each variation. Use k for the constant of proportionality.

1. y varies directly as n.

1. _____

2. x varies directly as r.

2. _____

Solve.

3. y varies directly as x. If $y = 20$ when $x = 5$, find y when x is 10.

3. _____

4. m varies directly as n. If $n = 343$ when $m = 49$, find m when n is 84.

4. _____

Objective 2

Write an equation to describe each variation. Use k for the constant of proportionality.

5. y varies inversely as t.

5. _____

Martin-Gay *Beginning Algebra, Fifth Edition*

6. s varies inversely as p.

6. _____

Solve.

7. y varies inversely as x. If $y = 12$ when $x = 10$, find y when x is 15.

7. _____

8. x varies inversely as w. If $x = 14$ when $w = 15$, find x when w is 10.

8. _____

Objective 3

Write an equation to describe each variation. Use k for the constant of proportionality.

9. x varies directly as the square of y.

9. _____

10. r varies directly as $\sqrt{t}$

10. _____

Solve.

11. a varies inversely as b^3. If $a = \dfrac{3}{2}$ when $b = 2$, find a when b is 3.

11. _____

12. y varies directly as x^2. If $y = 4$ when $x = 6$, find y when x is 10.

12. _____

Objective 4

13. The weight of a Kimodo dragon is directly proportional to its length. If an 8-foot Kimodo dragon weighs 200 pounds, then how much should a 6-foot Kimodo dragon weigh?

13. _____

14. The distance an object falls in a vacuum varies directly with the square of the time it is falling. If an object falls for 0.5 seconds, it will fall 4 feet. How far will an object fall in 3 seconds?

14. _____

15. The volume of a gas in a cylinder at a fixed temperature is inversely proportional to the weight on the piston. If the gas has a volume of 9 cubic centimeters for a weight of 20 kg, what would the volume be for a weight of 18 kg.

15. _____

Concept Extension

16. y varies directly as x^2.
        a. If x is doubled what is the effect on y?

16a. _____

        b. If x is quadrupled what is the effect on y?

16b. _____

Section 7.8 Simplifying Complex Fractions

Learning Objectives
1. Simplify complex fractions using method 1.
2. Simplify complex fractions using method 2.

Objective 1

Simplify these complex fractions using Method I.

1. $\dfrac{\dfrac{3}{n} - \dfrac{2}{m}}{\dfrac{7}{n} + \dfrac{1}{mn}}$

1. _____

2. $\dfrac{\dfrac{4}{x} + \dfrac{5}{x^2}}{\dfrac{3}{y^2} - \dfrac{7}{y}}$

2. _____

3. $\dfrac{\dfrac{1}{5} - \dfrac{1}{x}}{\dfrac{7}{10} + \dfrac{1}{x^2}}$

3. _____

4. $\dfrac{\dfrac{3}{x-2} - 4}{6 - \dfrac{1}{x+2}}$

4. _____

5. $2 - \dfrac{x}{3 - \dfrac{2}{x}}$

5. _____

Objective 2

Simplify these complex fractions by using Method II.

6. $\dfrac{5 - \dfrac{2}{x-3}}{4 - \dfrac{1}{x-3}}$

6. _____

7. $3 - \dfrac{\dfrac{2}{x-1}}{3 - \dfrac{4}{x-1}}$

7. _____

8. $\dfrac{\dfrac{2}{x+y} - \dfrac{3}{x-y}}{\dfrac{4}{x+y} + \dfrac{5}{x^2 - y^2}}$

8. _____

9. $\dfrac{\dfrac{x}{y} + 1}{\dfrac{x}{y} - 1}$

9. _____

Martin-Gay *Beginning Algebra, Fifth Edition*

10. $\dfrac{\dfrac{4}{x^2-4} + \dfrac{1}{x-2}}{\dfrac{6}{x+2} - \dfrac{4}{x-2}}$

10. _____

Concept Extension

Simplify the complex fraction.

11. $\dfrac{x^{-1} - 4^{-1}}{5^{-2} - x^{-2}}$

11. _____

Chapter 7 Vocabulary

Vocabulary Word	Definition	Example
Rational Expression	An expression that can be written as $\dfrac{P}{Q}$ where both P and Q are polynomials.	$\dfrac{x-1}{x+3}$
Multiply Rational Expressions	1. Factor numerators and denominators. 2. Multiply across 3. Simplify if needed.	$\dfrac{x-2}{x+3} \cdot \dfrac{2x}{x-1} = \dfrac{2x(x-2)}{(x+3)(x-1)}$
Divide Rational Expressions	Multiply by the reciprocal of the second rational expression.	$\dfrac{x-2}{x+3} \div \dfrac{2x}{x-1} = \dfrac{x-2}{x+3} \div \dfrac{x-1}{2x} =$ $\dfrac{2x(x-2)}{(x+3)(x-1)}$
Add or Subtract Rational Expressions	1. Factor the denominators. 2. Find an LCD. 3. Rewrite as equivalent fraction with the LCD. 4. Add or subtract. 5. Simplify if needed	$\dfrac{x-2}{x+3} + \dfrac{2x}{x-1} \quad LCD = (x+3)(x-1)$ $\dfrac{(x-2)(x-1)}{(x+3)(x-1)} + \dfrac{2x(x+3)}{(x+3)(x-1)} =$ $\dfrac{x^2-3x+2+2x^2+6x}{(x+3)(x-1)} = \dfrac{3x^2+3x+2}{(x+3)(x-1)}$
Ratio	Quotient of two numbers	$\dfrac{3}{5} = 3$ to 5
Proportion	A Mathematical statement that two ratios are equal.	$\dfrac{1}{4} = \dfrac{x}{16}$
Direct Variation	y varies directly as x	$y = kx$
Inverse Variation	y varies inversely as x	$y = \dfrac{k}{x}$

Martin-Gay *Beginning Algebra, Fifth Edition*

Simplify Complex Fraction -Method I	1. Add or subtract the numerator and the denominator. 2. Perform the division 3. Simplify if needed.	$\dfrac{\dfrac{1}{x}+\dfrac{1}{2}}{\dfrac{1}{x}-\dfrac{1}{4}}=\dfrac{\dfrac{2+x}{2x}}{\dfrac{4-x}{4x}}=$ $\dfrac{2+x}{2x}\cdot\dfrac{4x}{4-x}=\dfrac{4x(2+x)}{2x(4-x)}=\dfrac{2(2+x)}{4-x}$
Simplify Complex Fraction -Method II	1. Find LCD of all fractions. 2. Multiply numerator and denominator by that LCD. 3. Add/Subtract 4. Simplify if needed	$\dfrac{\dfrac{1}{x}+\dfrac{1}{2}}{\dfrac{1}{x}-\dfrac{1}{4}}\quad LCD=4x$ $\dfrac{4x\left(\dfrac{1}{x}+\dfrac{1}{2}\right)}{4x\left(\dfrac{1}{x}-\dfrac{1}{4}\right)}=\dfrac{4+2x}{4-x}=\dfrac{2(2+x)}{4-x}$

Name:

Instructor:

Practice Test A

Find any real numbers for which the following expression is undefined.

1. $\dfrac{x-3}{x^2-5x+6}$

1. _____

2. $\dfrac{x+3}{x^2-4}$

2. _____

3. $\dfrac{x-1}{x^2-6x}$

3. _____

Perform the indicated operation.

4. $\dfrac{x^2-3}{x+1}+\dfrac{2x-4}{x+1}$

4. _____

5. $\dfrac{x}{x-5}\cdot\dfrac{2}{x+4}$

5. _____

6. $\dfrac{x-2}{x^2+4x+4}\div\dfrac{2x+1}{x^2-4}$

6. _____

7. $\dfrac{x-3}{x^2-2x-15}-\dfrac{4-x}{x^2-9x+20}$

7. _____

8. $\dfrac{1}{x-1}+\dfrac{x}{1-x}$

8. _____

9. $\dfrac{3(x-4)}{x-1}\cdot\dfrac{x^2-1}{9x}$

9. _____

10. $\dfrac{x-2}{8x-24}\cdot\dfrac{5x-15}{x^2-4}$

10. _____

Simplify each complex fraction.

11. $\dfrac{2-\dfrac{1}{x}}{\dfrac{4}{x^2}-3}$

11. _____

12. $\dfrac{\dfrac{3}{x-1}+\dfrac{4x}{x+1}}{\dfrac{7}{x-1}-\dfrac{2}{x+1}}$

12. _____

Solve each equation.

13. $\dfrac{5}{x-1}=\dfrac{-2}{x+1}$

13. _____

14. $\dfrac{x}{x-1}=\dfrac{1}{2}+\dfrac{3}{x}$

14. _____

15. $\dfrac{3}{n+1} - \dfrac{1}{n+1} = \dfrac{14}{n^2-1}$

15. _____

Solve each equation for the indicated variable.

16. $m = \dfrac{r}{1+rt}$ for r.

16. _____

17. $b = \dfrac{2A-Bh}{h}$ for B

17. _____

Solve.

18. Hooke's law states that the distance a spring stretches is directly proportional to the weight attached to the spring. If a 50 pound weight stretches the spring 12 inches. How long would a 75 pound weight stretch the spring?

18. _____

19. It takes Joan 2 hours to fold the week's worth of laundry. It takes her husband 3 hours. If they work together, how long will it take them to fold the laundry?

19. _____

20. The sum of a number and 5 divided by 6 yields that number less 5, divided by 3. Find the number.

20. _____

Name:

Instructor:

Date:

Section:

Practice Test B

Find any real numbers for which the following expression is undefined.

1. $\dfrac{3x-2}{4x-1}$

 a. 0

 c. 1

 b. $\frac{1}{4}$

 d. $\frac{2}{3}$

1. _____

2. $\dfrac{x+1}{x^2-1}$

 a. 1 and -1

 c. 1

 b. 1

 d. -1

2. _____

3. $\dfrac{x+2}{x^2+3}$

 a. -3

 c. none

 b. -2

 d. all real numbers

3. _____

Perform the indicated operation.

4. $\dfrac{x+1}{x-1} - \dfrac{7}{x+1}$

 a. $\dfrac{x-6}{-2}$

 c. $\dfrac{x-6}{2x}$

 b. $\dfrac{x^2-5x-8}{x^2+1}$

 d. $\dfrac{x^2-5x+8}{x^2-1}$

4. _____

Martin-Gay *Beginning Algebra, Fifth Edition*

243

5. $\dfrac{x^2 + 16x + 64}{x+9} \div \dfrac{x + 17x + 72}{x+8}$

a. $\dfrac{(x+8)^2}{(x+9)^2}$

b. $\dfrac{(x+9)(x+8)}{x+8}$

5. _____

c. $\dfrac{x+9}{(x+8)^2}$

d. $\dfrac{(x+9)^2}{x+8}$

6. $5 - \dfrac{3x}{x-5}$

a. $2x - 5$

b. $\dfrac{2x - 25}{x-5}$

6. _____

c. $\dfrac{5x - 20}{x-5}$

d. $\dfrac{8x - 25}{x-5}$

7. $\dfrac{x-3}{x+1} - \dfrac{4-x}{x}$

a. $\dfrac{2x^2 - 6x - 4}{x(x+1)}$

b. $\dfrac{x^2 - 3x + 4}{x(x+1)}$

7. _____

c. $\dfrac{2x^2 - 6x + 4}{x(x-1)}$

d. $\dfrac{x^2 - 4x - 4}{x(x+1)}$

8. $\dfrac{x+3}{x^2 + 2x - 8} \cdot \dfrac{x - 2x}{x^2 + 6x + 9}$

a. $\dfrac{x+3}{x-2}$

b. $\dfrac{(x+3)(x+4)}{(x-2)}$

8. _____

c. $\dfrac{x}{(x+4)(x+3)}$

d. $\dfrac{x(x+3)}{(x-2)(x-3)}$

9. $\dfrac{x+4}{x^2+1} \div \dfrac{x^2-16}{x+1}$

   a. $\dfrac{x+4}{x^2+1}$

   b. $\dfrac{x}{(x+4)(x+3)}$

   c. $\dfrac{x^2+1}{(x+3)(x+4)}$

   d. $\dfrac{x+3}{(x+4)(x^2+1)}$

9. _____

10. $\dfrac{x+1}{x^2+9} + \dfrac{x}{x-3}$

   a. $\dfrac{x^2-9}{(x+1)(x+4)}$

   b. $\dfrac{x^2+1}{(x-3)(x+3)}$

   c. $\dfrac{x^2+4x+1}{x^2-9}$

   d. $\dfrac{(x-3)(x+1)}{x(x+3)}$

10. _____

Simplify each complex fraction.

11. $\dfrac{\dfrac{1}{x}+\dfrac{3}{4x}}{\dfrac{x}{6}-\dfrac{3}{x^2}}$

   a. $\dfrac{2x^3-36}{21x}$

   b. $\dfrac{3x}{2(x-18)}$

   c. $\dfrac{21x}{2x^3-36}$

   d. $\dfrac{18x}{2x^2-36}$

11. _____

12. $\dfrac{\dfrac{3}{x-1}+\dfrac{4x}{x+1}}{\dfrac{7}{x-1}-\dfrac{2}{x+1}}$

   a. $\dfrac{4x^2+7x-3}{9x+9}$

   b. $\dfrac{5x^2-3x+3}{9x+5}$

12. _____

c. $\dfrac{7x^2 - 4x + 3}{5x + 9}$ d. $\dfrac{4x^2 - x + 3}{5x + 9}$

Solve each equation.

13. $\dfrac{4}{x-5} - \dfrac{3}{x+2} = \dfrac{28}{x^2 - 3x - 10}$

13. _____

    a. 5   b. - 5

    c. -2   d. no solution

14. $\dfrac{x}{x^2 - 9} + \dfrac{4}{4x - 12} = \dfrac{-3}{x}$

14. _____

    a. 1   b. $^{17}\!/_{33}$

    c. $^{33}\!/_{17}$ d. no solution

15. $\dfrac{5-x}{5+x} = \dfrac{6}{7}$

15. _____

    a. -5   b. $^{5}\!/_{13}$

    c. -1 d. no solution

Solve each equation for the indicated variable.

16. $\dfrac{y-b}{x} = m$ for b.

    a. $b = \dfrac{y-m}{x}$ b. $b = \dfrac{y}{mx}$

16. _____

c. $b = \dfrac{x - m}{y}$ d. $m = y - mx$

Solve.

17. y varies inversely as x. If $x = 40$ when $y = 240$, find y when $x = 160$.

17. _____

   a. 60   b. 6

   c. 960   d. 360

18. The distance a car travels is directly proportional to the time it is traveling. If a car can travel 200 miles in 4 hours, how long will it take the car to travel 350 miles?

18. _____

   a. 6 hours   b. 5 hours

   c. 8 hours   d. 7 hours

19. Given these are similar triangles, find the measure of x.

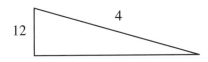   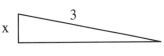

19. _____

   a. 9   b. 1

   c. 6   d. 8

20. It takes a hose 20 hours to fill a pool. Another hose can fill the pool in 15 hours. Approximately how long will it take the pool to fill if both hoses are used?

20. _____

   a. 9.2 hours   b. 8.6 hours

   c. 10.4 hours   d. 9.8 hours

Chapter 8 Roots and Radicals
Section 8.1 Introduction to Radicals

Learning Objectives
1. Find square roots.
2. Find cube roots.
3. Find *n*th roots.
4. Approximate square roots.
5. Simplify radicals containing variables.

Vocabulary
Use the choices to complete each statement.
Index   **Perfect**   **Principal**
Radical sign   **Radicand**

1. In the expression $\sqrt[4]{6}$, the number 6 is called the _____ and the number 4 is called the

   _____ .

2. In the expression $\sqrt{5}$, the $\sqrt{}$ is called the _____ . It symbolizes the _____

   square root of the 5.

3. A number is a _____ square if the square root of it can be simplified to rational numbers.

Objective 1

Find each square root.

4. $\sqrt{49}$

4. _____

5. $-\sqrt{121}$

5. _____

6. $\sqrt{-4}$

6. _____

7. $\sqrt{\dfrac{4}{25}}$

7. _____

8. $\sqrt{0.064}$

8. _____

Objective 2

Find each cube root.

9. $\sqrt[3]{64}$

9. _____

10. $\sqrt[3]{-27}$

10. _____

11. $\sqrt[3]{\dfrac{1}{125}}$

11. _____

12. $\sqrt[3]{-0.000512}$

12. _____

13. $\sqrt[3]{\dfrac{216}{343}}$

13. _____

Objective 3

Find each root.

14. $\sqrt[5]{-1024}$

14. _____

15. $-\sqrt[4]{4096}$

15. _____

16. $\sqrt[6]{-1}$

16. _____

17. $\sqrt[5]{\dfrac{1}{32}}$

17. _____

Objective 4

Approximate each square root to three decimal places.

18. $\sqrt{56}$

18. _____

19. $\sqrt{241}$

19. _____

Martin-Gay *Beginning Algebra, Fifth Edition*

20. $\sqrt{0.45}$

20. _____

21. $\sqrt{44}$

21. _____

Objective 5

Find each root. Assume that all variables represent positive numbers.

 22. $\sqrt{36x^{12}}$

22. _____

23. $\sqrt{81y^6z^8}$

23. _____

24. $-\sqrt[3]{-64n^9m^{12}}$

24. _____

25. $\sqrt[4]{x^{16}y^{20}z^8}$

25. _____

Concept Extension

Recall from this section that $\sqrt{a^2} = |a|$ for any real number a. Simplify the following given that x represents any real number.

 26. $\sqrt{x^2}$

26. _____

27. $\sqrt{x^2 + 10x + 25}$

27. _____

28. $\sqrt{36x^2 + 84x + 49}$

28. _____

Name: Date:

Instructor: Section:

Section 8.2 Simplifying Radicals

Learning Objectives
1. Use the product rule to simplify square roots.
2. Use the quotient rule to simplify square roots.
3. Simplify radicals containing variables.
4. Simplify higher roots.

Objective 1

Use the product rule to simplify each radical.

1. $\sqrt{180}$

1. _____

2. $\sqrt{250}$

2. _____

3. $-3\sqrt{32}$

3. _____

4. $5\sqrt{64}$

4. _____

5. $\sqrt{200}$

5. _____

Objective 2

Use the quotient rule and the product rule to simplify each radical.

 6. $\sqrt{\dfrac{27}{121}}$

6. _____

7. $-\sqrt{\dfrac{45}{64}}$

7. _____

8. $\sqrt{\dfrac{28}{81}}$

8. _____

9. $-\sqrt{\dfrac{120}{144}}$

9. _____

Objective 3

Simplify each radical. Assume that all variables represent positive numbers.

 10. $\sqrt{81b^5}$

10. _____

11. $\sqrt{32t^6}$

11. _____

12. $\sqrt{\dfrac{x^7}{25y^4}}$

12. _____

13. $\sqrt{\dfrac{250m^3}{9n^6}}$

13. _____

Objective 4

Simplify.

 14. $\sqrt[4]{\dfrac{8}{81}}$

14. _____

15. $\sqrt[5]{160}$

15. _____

16. $\sqrt[3]{\dfrac{688}{1000}}$

16. _____

17. $\sqrt[4]{-81}$

17. _____

Concept Extension

18. Determine if $\sqrt{a^2 + b^2} = a + b$ for all real numbers.

18. _____

19. The time t in seconds it takes an object to fall d feet is given by the equation. $t = \dfrac{1}{4}\sqrt{d}$.

  a. The height of the Eiffel Tower in Paris is 1,063 feet tall. How long would it take a feather to fall to the ground from the top of the Eiffel Tower.

19. _____

Section 8.3 Adding and Subtracting Radicals.

Learning Objectives
1. Add or subtract like radicals.
2. Simplify radical expressions, and then add or subtract any like radicals.

Objective 1

Simplify each expression by combining like radicals where possible.

1. $5\sqrt{2} - 7\sqrt{2}$

1. _____

 2. $3\sqrt{6} + 8\sqrt{6} - 2\sqrt{6} - 5$

2. _____

3. $7\sqrt[3]{6} - 10\sqrt[3]{6} + 4\sqrt{6}$

3. _____

4. $14\sqrt{3} + 16 - 3\sqrt{3} + 4 + 5\sqrt{3} - 5$

4. _____

5. $\sqrt{10} - 5\sqrt{10} + 8\sqrt{10}$

5. _____

Objective 2

Add or subtract by first simplifying each radical and them combining any like radical terms. Assume that all variables represent positive real numbers.

 6. $5\sqrt{2x} + \sqrt{98x}$

6. _____

7. $5x\sqrt{8} + 3\sqrt{32x^2} - 5\sqrt{50x^2}$

7. _____

8. $7\sqrt{75nm^3} - 4m\sqrt{12nm}$

8. _____

Martin-Gay *Beginning Algebra, Fifth Edition*

9. $\sqrt{\dfrac{3}{64}} + \sqrt{\dfrac{3}{16}}$

9. _____

10. $13\sqrt[3]{x^4 y^3 z^2} - 15xy\sqrt[3]{xz^2}$

10. _____

11. $t\sqrt[4]{5tu^8} + u\sqrt[4]{405t^5u^4} + u^2\sqrt[4]{80t^5}$

11. _____

Concept Extension

Add or subtract by first simplifying each radical and them combining any like radical terms. Assume that all variables represent positive real numbers.

12. $\sqrt{\dfrac{y^5}{16}} + y\sqrt{\dfrac{4y^3}{25}} - \dfrac{y^2\sqrt{144}}{2}$

12. _____

Section 8.4 Multiplying and Dividing Radicals

Learning Objectives
1. Multiply radicals.
2. Divide radicals.
3. Rationalize denominators.
4. Rationalize using conjugates.

Objective 1

Multiply and simplify. Assume that all the variables represent positive real numbers.

1. $\sqrt{32} \cdot \sqrt{20}$

1. _____

2. $\sqrt{64x^3} \cdot \sqrt{16x}$

2. _____

3. $\sqrt{3}\left(\sqrt{2} - \sqrt{5}\right)$

3. _____

4. $\sqrt{6y}\left(\sqrt{2y^3} + \sqrt{3y}\right)$

4. _____

5. $\left(\sqrt{7y} - \sqrt{3x}\right)\left(\sqrt{21y} + \sqrt{6x}\right)$

5. _____

6. $\left(\sqrt{x} + 6\right)\left(\sqrt{x} - 6\right)$

6. _____

7. $\left(\sqrt{2} + \sqrt{5x}\right)^2$

7. _____

Martin-Gay *Beginning Algebra, Fifth Edition*

Objective 2

Divide and simplify. Assume that all the variables represent positive real numbers.

8. $\dfrac{\sqrt{280}}{\sqrt{10}}$

8. _____

9. $\dfrac{\sqrt{125y^3}}{\sqrt{5y}}$

9. _____

10. $\dfrac{a\sqrt{360a^5}}{\sqrt{6a}}$

10. _____

11. $\dfrac{-\sqrt{864x^5}}{\sqrt{8x^3}}$

11. _____

Objective 3

Rationalize each denominator and simplify. Assume that all the variables represent positive real numbers.

12. $\dfrac{4}{\sqrt{5}}$

12. _____

13. $\dfrac{\sqrt{3}}{\sqrt{6a}}$

13. _____

14. $\dfrac{7}{\sqrt{18c}}$

14. _____

15. $\sqrt{\dfrac{4m}{7n}}$

15. _____

16. $\sqrt{\dfrac{1}{48}}$

16. _____

17. $\sqrt[3]{\dfrac{1}{9}}$

17. _____

Objective 4

Rationalize each denominator and simplify. Assume that all the variables represent positive real numbers.

 18. $\dfrac{4}{2-\sqrt{5}}$

18. _____

19. $\dfrac{7}{3-\sqrt{x}}$

19. _____

20. $\dfrac{\sqrt{2}}{4-\sqrt{3}}$

20. _____

21. $\dfrac{6-\sqrt{5}}{4-\sqrt{3}}$

21. _____

22. $\dfrac{3+\sqrt{x}}{2-\sqrt{x}}$

22. _____

Concept Extension

Rationalize the numerator of the following.

23. $\dfrac{\sqrt{5}}{4}$

23. _____

24. $\dfrac{2-\sqrt{5}}{3+\sqrt{3}}$

24. _____

Section 8.5 Solving Equations Containing Radicals

Learning Objectives
  1. Solve radical equations by using the squaring property of equality once.
  2. Solve radical equations by using the squaring property of equality twice.

Objective 1

Solve each equation.

1. $\sqrt{x-4} = 6$

1. _____

2. $3\sqrt{x} - 5 = 11$

2. _____

3. $\sqrt{2x-5} = \sqrt{4x+1}$

3. _____

4. $\sqrt{x^2 + 6x + 18} = 9$

4. _____

5. $\sqrt{1-8x} - x = 4$

5. _____

6. $\sqrt{5x+4} = \sqrt{2x-8}$

6. _____

Martin-Gay *Beginning Algebra, Fifth Edition*

7. $\sqrt{a+3} = a-3$

8. $\sqrt{4x+5} + 3 = 6$

Objective 2

Solve each equation.

9. $\sqrt{x-7} = \sqrt{x}+1$

10. $\sqrt{y+8} - \sqrt{y-4} = 2$

11. $\sqrt{5x} - 1 = \sqrt{5x+1}$

12. $\sqrt{2t-1} - 2 = \sqrt{t-4}$

13. $\sqrt{6-y} - \sqrt{y-2} = 2$

14. $\sqrt{3a} - \sqrt{a-2} = 4$

14. _____

Concept Extension

Solve each equation.

15. $\sqrt[3]{x-2} = 3$

15. _____

Section 8.6 Radical Equations and Problem Solving

Learning Objectives
1. Use the Pythagorean formula to solve problems.
2. Use the distance formula.
3. Solve problems using formulas containing radicals.

Objective 1

Use the Pythagorean theorem to find the unknown side of each right triangle.

1.

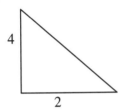

1. _____

2.

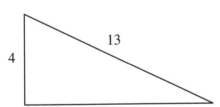

2. _____

3.

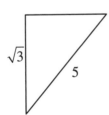

3. _____

Find the length of each unknown side of each right triangle with sides a, b, and c, where c is the hypotenuse.

4. $a = 3, b = 15$

4. _____

5. $c = \sqrt{50},\ a = \sqrt{10}$

5. _____

6. $a = 6,\ c = 20$

6. _____

Objective 2

Use the distance formula to find the distance between the points given.

7. $(1,2),\ (6,8)$

7. _____

 8. $(-3,1),(5,-2)$

8. _____

9. $\left(\dfrac{1}{2},1\right),\left(-2,\dfrac{3}{4}\right)$

9. _____

10. $(4,-5),(2,6)$

10. _____

Name:

Instructor:

Date:

Section:

Objective 3

Solve each problem.

11. A wire is used to anchor a 20-foot high pole. One end of the wire is attached to the top of the pole. The other end is fastened to a stake five feet away from the bottom of the pole. Find the length of the wire, to the nearest tenth of a foot.

11. _____

12. One end of a glass prism is in the shape of a right triangle with a hypotenuse of 20 inches, and a height of 10 inches.
   a. Find the measurement of the base of the triangle.

12a. _____

   b. What is the area of the triangle?

12b. _____

13. Joe traveled 6 miles east, then 12 miles south. How far is he away from the point he originally started it.

13. _____

14. Prove if this triangle is a right triangle.

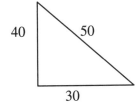

14. _____

Concept Extension

15. Find the distance between these two points. $(4.5, 1.3), (2.0, -2.3)$

15. _____

16. Using the figure below, answer the following questions. Each side of the cube measures 2 inches.

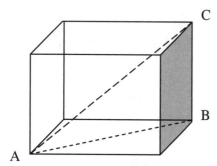

a. Find the length of the line connecting points A and B.

16a. _____

b. Find the length on the line connecting points A and C.

16b. _____

Name:

Instructor:

Date:

Section:

Section 8.7 Rational Exponents

Learning Objectives

1. Evaluate exponential expressions of the form $a^{1/n}$

2. Evaluate exponential expressions of the form $a^{m/n}$

3. Evaluate exponential expressions of the form $a^{-m/n}$

4. Use the rules of exponents to simplify expressions containing fractional exponents.

Objective 1

Simplify each expression.

1. $36^{1/2}$

1. _____

2. $-32^{1/5}$

2. _____

3. $(0.25)^{1/2}$

3. _____

4. $\left(\dfrac{1}{16}\right)^{1/4}$

4. _____

5. $-(81)^{1/2}$

5. _____

Objective 2

Simplify each expression.

6. $4^{3/2}$

6. _____

7. $8^{2/3}$

7. _____

8. $\left(\dfrac{-1}{27}\right)^{4/3}$

8. _____

9. $32^{4/5}$

9. _____

10. $-\left(\dfrac{27}{64}\right)^{2/3}$

10. _____

Objective 3

Simplify each expression.

11. $-16^{-1/4}$

11. _____

12. $343^{-2/3}$

12. _____

13. $\left(\dfrac{64}{125}\right)^{-1/3}$

13. _____

14. $(-8)^{-\frac{2}{3}}$

14. _____

Objective 4

Simplify each expression. Write each answer with positive exponents. Assume all the variables represent positive numbers.

15. $2^{\frac{1}{3}} \cdot 2^{\frac{1}{2}}$

15. _____

16. $\dfrac{3^{-\frac{3}{5}}}{3^{\frac{2}{5}}}$

16. _____

17. $\left(x^{\frac{3}{5}}\right)^{5} \cdot x^{2}$

17. _____

18. $\dfrac{\left(n^{\frac{2}{3}}\right)^{4}}{\left(n^{\frac{1}{2}}\right)^{3}}$

18. _____

19. $\left(2^{\frac{1}{2}}\right)^{-\frac{2}{3}}$

19. _____

Concept Extension

Solve.

20. $(x+2)^{\frac{1}{2}} = 13$

20. _____

21. $(y+7)^{\frac{1}{2}} - 1 = (y+2)^{\frac{1}{2}}$

21. _____

Chapter 8 Vocabulary

Vocabulary Word	Definition	Example
Principal Square Root	Positive square root	The principal square root of 5 is $\sqrt{5}$.
Cube Root	$\sqrt[3]{a} = b$ $b^3 = a$	$\sqrt[3]{27} = 3$
nth root	$\sqrt[n]{a}$	$\sqrt[5]{32} = 2$
Product Rule for Radicals	$\sqrt[n]{a} \cdot \sqrt[n]{b} = \sqrt[n]{ab}$	$\sqrt[3]{2} \cdot \sqrt[3]{6} = \sqrt[3]{12}$
Quotient Rule for Radicals	$\dfrac{\sqrt[n]{a}}{\sqrt[n]{b}} = \sqrt[n]{\dfrac{a}{b}}$	$\dfrac{\sqrt[3]{16}}{\sqrt[3]{4}} = \sqrt[3]{\dfrac{16}{4}} = \sqrt[3]{4}$
Like Radicals	Radical expressions that have the same index and the same radicand.	$\sqrt{5}, 3\sqrt{5}, -6\sqrt{5}$ are like radicals.
Rationalize a denominator	Process of removing the radicals from the denominator of a radical expression.	$\dfrac{4}{\sqrt{2}} = \dfrac{4}{\sqrt{2}} \cdot \dfrac{\sqrt{2}}{\sqrt{2}} = \dfrac{4\sqrt{2}}{2} = 2\sqrt{2}$
Conjugate	The conjugate of $a+b$ is $a-b$	$5-\sqrt{3}$ and $5+\sqrt{3}$ are conjugates.
Solving radical equations.	1. Isolate one radical to one side. 2. Square both sides. 3. Simplify both sides. 4. Repeat steps 1 – 3, if there is another radical. 5. Solve the equation.	$\sqrt{x-4} = 5$ $x-4 = 5^2$ $x-4 = 25$ $x = 29$
Rational Exponents	$a^{1/n} = \sqrt[n]{a}$ $a^{m/n} = \sqrt[n]{a^m}$ $a^{-m/n} = \dfrac{1}{a^{m/n}}$	$2^{1/3} = \sqrt[3]{2}$ $2^{2/3} = \sqrt[3]{2^2}$ $2^{-2/3} = \dfrac{1}{2^{2/3}}$

Practice Test A

Simplify each expression. Assume the variables represent positive real numbers.

1. $\sqrt{125x^4y^6}$

1. _____

2. $\left(x^8y^{12}z^{18}\right)^{-\frac{1}{2}}$

2. _____

3. $\dfrac{\sqrt{18x}}{\sqrt{2}}$

3. _____

4. $\sqrt{3}\left(1-\sqrt{5}\right)$

4. _____

5. $\sqrt[4]{80u^6v^5}$

5. _____

Add or subtract by simplifying each radical and then combining like radicals.

6. $\sqrt{72}+\sqrt{50}-\sqrt{12}$

6. _____

7. $\sqrt{18x^3}-x\sqrt{98x}+7\sqrt{32}$

7. _____

Martin-Gay *Beginning Algebra, Fifth Edition*

Rationalize the denominator.

8. $\dfrac{4}{5-\sqrt{7}}$

8. _____

9. $\dfrac{\sqrt{6}-\sqrt{2}}{\sqrt{3}+\sqrt{5}}$

9. _____

10. $\dfrac{9}{10^{1/2}}$

10. _____

Solve each radical equation.

11. $\sqrt{x-11}=\sqrt{x+10}$

11. _____

12. $\sqrt{x+4}-2\sqrt{x-1}=-1$

12. _____

13. $2\sqrt{x-7}=3$

13. _____

14. $\sqrt{x^2 + 3x + 6} = 4$

14. _____

Solve.

15. Find the length of the unknown leg of a right triangle if the hypotenuse is 15 inches and one of the leg measures to be 12 inches.

15. _____

16. Find the distance between these two points $(5, -7), (-2, 4)$

16. _____

17. The radius of a circle can found if given the area by using the formula. $r = \sqrt{\dfrac{A}{\pi}}$. Find the radius of the circle if the area is 60 square inches. For calculation purposes we will say that $\pi \approx 3$.

17. _____

18. Find the length to a side of a square whose area is 72 square feet.

18. _____

Simplify each expression using positive exponents only.

19. $\dfrac{5^{4/5}}{5^{-2/5}}$

19. _____

20. $7^{1/2} \left(7^{-3/4} \right)$

20. _____

Practice Test B

Simplify each expression. Assume the variables represent positive real numbers.

1. $\sqrt{64x^6 y^{10}}$

    1. _____

    a. $8x^3 y^5$ b. $4x^3 y^5$

    c. $8x^4 y^8$ d. $4x^4 y^8$

2. $\left(a^9 b^{15} c^{12}\right)^{2/3}$

    2. _____

    a. $a^8 b^{12} c^{18}$ b. $a^{18} b^{10} c^8$

    c. $a^5 b^7 c^6$ d. $a^{20} b^{12} c^{18}$

3. $\dfrac{\sqrt{125x}}{\sqrt{5}}$

    3. _____

    a. $5x$ b. $\sqrt{5x}$

    c. $5\sqrt{x}$ d. $x\sqrt{5}$

4. $\sqrt{7}\left(\sqrt{3}+\sqrt{8}\right)$

    4. _____

    a. $\sqrt{21}+2\sqrt{14}$ b. $\sqrt{21}-\sqrt{15}$

    c. $\sqrt{10}+\sqrt{15}$ d. $\sqrt{21}+\sqrt{56}$

5. $\sqrt[5]{-32x^{10} y^{11}}$

    5. _____

    a. $-2x^2 y^2 \sqrt[5]{y}$ b. $2xy^2 \sqrt[5]{y}$

    c. $-2y^2 \sqrt[5]{xy}$ d. $2xy\sqrt[5]{y}$

Add or subtract by simplifying each radical and then combining like radicals.

6. $\sqrt{48} + 4\sqrt{75} - \sqrt{108}$

6. _____

    a. $16\sqrt{3}$ b. $18\sqrt{3}$

    c. $5\sqrt{3} - 3\sqrt{2}$ d. $32\sqrt{2}$

7. $\sqrt{80x} - \sqrt{20x^3} + 4x\sqrt{5x}$

7. _____

    a. $5x\sqrt{5x}$ b. $4\sqrt{5x} - 2\sqrt{5}$

    c. $3\sqrt{5x} - 4\sqrt{5}$ d. $(4 + 2x)\sqrt{5x}$

Rationalize the denominator.

8. $\dfrac{x}{4 - \sqrt{x}}$

8. _____

    a. $\dfrac{4x + x\sqrt{x}}{4 - x}$ b. $\dfrac{4 + 4x\sqrt{x}}{4 + x}$

    c. $\dfrac{4x\sqrt{x}}{x + 4}$ d. $\dfrac{4 - \sqrt{x}}{4 + x}$

9. $\dfrac{\sqrt{7} + \sqrt{10}}{\sqrt{3x} - \sqrt{12}}$

9. _____

    a. $\dfrac{\sqrt{21x} + 3x\sqrt{50}}{3x - 4}$ b. $\dfrac{\sqrt{21x} + 2\sqrt{21} + 2\sqrt{3} + 2\sqrt{30}}{3x - 12}$

    c. $\dfrac{\sqrt{21x} + 4\sqrt{3} + 2\sqrt{30}}{3x - 12}$ d. $\dfrac{2x\sqrt{30} - \sqrt{21x}}{3x - 4}$

10. $\dfrac{5}{6^{1/2}}$

10. _____

    a. $\dfrac{5\sqrt{6}}{6}$ b. $\dfrac{\sqrt{6}}{5}$

    c. $\dfrac{\sqrt{5}}{\sqrt{6}}$ d. $\sqrt{30}$

Solve each radical equation.

11. $\sqrt{2x+3} = \sqrt{5x-17}$

11. _____

    a. $20/3$ b. 20

    c. −20   d. no solution

12. $\sqrt{5x^2 - 9} = 2x$

12. _____

    a. 0 and -4   b. 2 and -4

    c. 3 and -3   d. 3 and 4

13. $\sqrt{5x - x^2} = \sqrt{6}$

13. _____

    a. 3 and 2   b. 1 and -6

    c. 3 and − 3   d. 4 and 5

14. $\sqrt{y} - \sqrt{y-1} = 1$

      a. -1 and 0 b. 1

      c. 1 and -1 d. no solution

14. _____

15. $\sqrt{x-3} + 1 = \sqrt{x-2}$

      a. 2 b. 2 and -3

      c. 3 d. no solution

15. _____

Solve.

16. Find the distance between these two points $(-6, -9), (-1, 2)$

16. _____

      a. $\sqrt{138}$ b. $\sqrt{146}$

      c. 12 d. $\sqrt{140}$

17. A 16 foot ladder is laid on a brick wall. The bottom of the ladder is 6 feet away from the wall. How high up the wall is the top of the ladder?

17. _____

      a. $2\sqrt{15}$ ft   b. $2\sqrt{55}$ ft

      c. $\sqrt{10}$ ft.   d. 12 ft

18. A foot ball player scored a touchdown by running along the diagonal of the 300 foot football field. How long did the player run if a football field is 160 feet wide.

18. _____

    a. 112 yards   b. 300 ft

    c. $50\sqrt{3}$ ft   d. 340 ft

Simplify each expression using positive exponents only.

19. $\left(7^{2/3}\right)^{4/5}$

19. _____

    a. $7^{5/6}$ b. $7^{3/4}$

    c. $7^{8/15}$ d. $7^{22/15}$

20. $\dfrac{3^{1/3}}{3^{1/2}}$

20. _____

    a. $3^{2/5}$ b. $\dfrac{1}{3^{1/6}}$

    c. $3^{5/6}$ d. $3^{1/6}$

Chapter 9 Quadratic Equations
Section 9.1 Solving Quadratic Equations by the Square Root Property

> **Learning Objectives**
>   1. Use the square root property to solve quadratic equations.
>   2. Solve problems modeled by quadratic equations.

Objective 1

Use the square root property to solve each quadratic equation.

1. $x^2 = 144$

1. _____

2. $x^2 + 4 = 40$

2. _____

3. $5x^2 = 125$

3. _____

4. $x^2 = -4$

4. _____

5. $(x-3)^2 = 25$

5. _____

6. $(13-3x)^2 = 169$

6. _____

7. $(p+2)^2 = 10$

7. _____

8. $(3x+5)^2 = 72$

8. _____

9. $\dfrac{2}{3}x^2 = 15$

9. _____

10. $(5x+2)^2 - 50 = -25$

10. _____

Objective 2

Solve.

11. Find two positive real numbers that have a sum of 8 and a product of 4

11. _____

12. The diagonal of a square is 3 meters longer than a side. Find the length of a side.

12. _____

13. The area of a circle is 288π square units. Find the radius.

13. _____

14. The sides of a right triangle are three consecutive even integers. Find the lengths of the three sides.

14. _____

Concept Extensions

Solve each quadratic function by first factoring the perfect square trinomial on the left side.

15. $4x^2 - 24x + 9 = 21$

15. _____

16. $y^2 - 2.4y + 1.44 = 1.69$

16. _____

Section 9.2 Solving Quadratic Equations by Completing the Square

Learning Objectives
1. Write perfect square trinomials.
2. Solve quadratic equations of the form $x^2 + bx + c = 0$ by completing the squares.
3. Solve quadratic equations of the form $ax^2 + bx + c = 0$ by completing the squares.

Objective 1

Complete the square for each expression and then factor the resulting perfect square trinomial.

1. $x^2 - 2x$

1. _____

2. $x^2 + 10x$

2. _____

3. $x^2 + 7x$

3. _____

4. $x^2 - 9x$

4. _____

Objective 2

Solve each quadratic equation by completing the square.

 5. $x^2 + 8x = -12$

5. _____

6. $n^2 - 8n + 4 = 0$

6. _____

7. $r^2 + r = 0$

7. _____

8. $x(x-6) = 12$

8. _____

9. $t^2 - 3t + 5 = 0$

9. _____

Objective 2

Solve each quadratic equation by completing the square.

 10. $2y^2 + 8y + 5 = 0$

10. _____

11. $3x^2 - 6x - 16 = 0$

11. _____

12. $5t^2 + 15t = 20$

12. _____

13. $4x^2 - 5x - 1 = 0$

13. _____

14. $2p^2 + p - 6 = 0$

14. _____

Concept Extension

15. Find the values for b that will make $x^2 + bx + 64$ a perfect square trinomial.

15. _____

16. Find both the x-intercepts and y-intercepts of the equation $y = x^2 - 8x$ by completing the square.

16. _____

Section 9.3 Solving Quadratic Equations by the Quadratic Formula

Learning Objective
1. Use the quadratic formula to solve quadratic equations.
2. Approximate solutions to quadratic equation.
3. Determine the number of solutions of a quadratic equation by using the discriminant.

Objective 1

Use the quadratic formula to solve each quadratic equation.

1. $3k^2 + 7k + 1 = 0$

1. _____

2. $2x^2 - 4x + 5 = 0$

2. _____

3. $5x^2 - 4x - 11 = 0$

3. _____

4. $5 - 2x - 6x^2 = 0$

4. _____

5. $x^2 + 5x - 4 = 0$

5. _____

6. $3x^2 + 6x - 2 = 1$

6. _____

Objective 2

Approximate the solution to these quadratic equations to three decimal places by using the quadratic equation.

7. $x^2 - 7x + 13 = 0$

7. _____

8. $x^2 = 9x + 4$

8. _____

9. $3x^2 - 4x - 2 = 0$

9. _____

10. $2x^2 - 10x + 8 = 0$

10. _____

11. $5x^2 = 13$

11. _____

Objective 3

Use the discriminant to determine the number of solutions of each quadratic equation.

 12. $4x^2 + 4x = -1$

12. _____

13. $3x^2 - 5x + 6 = 0$

13. _____

14. $5x^2 - 6x - 8 = 0$

14. _____

15. $x^2 - 4x + 11 = 0$

15. _____

16. $17 - 14x - 3x^2 = 0$

16. _____

Concept Extension

Solve. Approximate your answer to three decimal places.

17. $1.2x^2 + 4.2x = 3.4$

17. _____

Section 9.4 Complex Solutions of Quadratic Equations

> **Learning Objectives**
> 1. Write complex numbers using i notation.
> 2. Add and subtract complex numbers.
> 3. Multiply complex numbers.
> 4. Divide complex solutions.
> 5. Solve quadratic equations that have complex solutions.

Vocabulary
Use the choices to complete each statement.

Complex   **Conjugate**
Imaginary   **Real**

1. The result of $\sqrt{-25}$ will be a(n) _____ number.

2. $3 + 5i$ is an example of a(n) _____ number.

3. The _____ of $2 - 4i$ is $2 + 4i$.

4. $5 - 0i$ is an example of a _____ number.

Objective 1

Write each expression in i notation.

5. $\sqrt{-36}$

5. _____

6. $\sqrt{-121}$

6. _____

7. $\sqrt[4]{-50}$

7. _____

8. $\sqrt{-56x^2}$

8. _____

Objective 2

Add or subtract as indicated.

9. $(4 + i) + (-6 - 5i)$

9. _____

Martin-Gay *Beginning Algebra, Fifth Edition*   289

10. $(5+8i)-(8-3i)$

10. _____

11. $(2+7i)-(11-9i)$

11. _____

12. $(13+17i)+(18-14i)$

12. _____

Objective 3

Multiply.

13. $-5i(3-2i)$

13. _____

14. $(4+i)(8-3i)$

14. _____

15. $(4-3i)(4+3i)$

15. _____

16. $(4-8i)^2$

16. _____

17. $(3-7i)(-3-8i)$

17. _____

Martin-Gay *Beginning Algebra, Fifth Edition*

Objective 4

Divide. Write each answer in standard form.

 18. $\dfrac{7-i}{4-3i}$

18. _____

19. $\dfrac{2+i}{5+2i}$

19. _____

20. $\dfrac{6-2i}{3+9i}$

20. _____

21. $\dfrac{6-5i}{i}$

21. _____

Objective 5

Solve the following quadratic equations for complex solutions.

22. $(x+1)^2 = -9$

22. _____

23. $y^2 + 6y = 3$

23. _____

24. $3n^2 + 7 = 8n$

24. _____

25. $4x^2 - 3x + 5 = 0$

25. _____

26. $7t^2 = 5t - 2$

26. _____

Concept Extension

Solve the equation.

27. $(x-4)^2 + (x+2)(x+1) = 9$

27. _____

Section 9.5 Graphing Quadratic Equations

Learning Objective

    1. Graph quadratic equations of the form ax^2

    2. Graph quadratic equations of the form $ax^2 + bx + c$

    3. Use the vertex formula to determine the vertex of a parabola.

Objective 1

Graph each quadratic equation by finding and plotting ordered pairs solutions.

    1. $y = 2x^2$

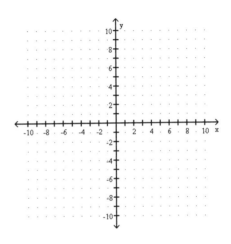

    2. $y = -3x^2$

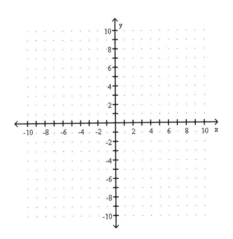

3. $y = \dfrac{1}{3}x^2$

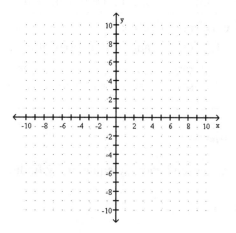

4. $y = -\dfrac{1}{4}x^2$

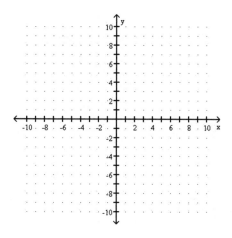

5. $y = 1.25x^2$

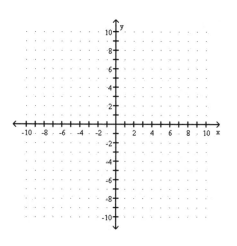

Objective 2

Sketch the graph of each equation.

6. $y = 2x^2 - 11x + 5$

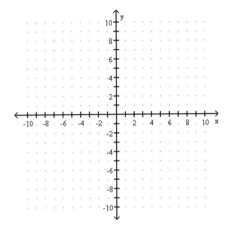

7. $y = -x^2 - 4x - 3$

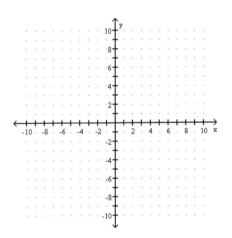

8. $y = 9x^2 - 6x + 1$

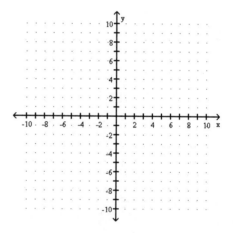

9. $y = 3x^2 - 7x + 2$

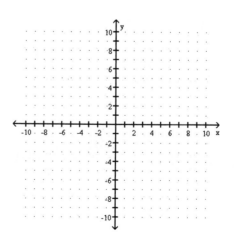

10. $y = 8x^2 + 2x - 1$

Objective 3

Use the vertex formula to find the coordinate of the vertex for the given quadratic equation.

11. $y = x^2 - 4x + 9$

11. _____

12. $y = 2x^2 - 11x + 20$

12. _____

13. $y = -3x^2 + 5x - 10$

13. _____

14. $y = -x^2 + x - 3$

14. _____

Concept Extension

15. Determine the approximate vertex and x-intercepts for the quadratic equation $y = 1.6x^2 - 3.2x + 2.4$

15. _____

16. Graph $y = 1.6x^2 - 3.2x + 2.4$, what is the domain and range of this quadratic equation.

16. _____

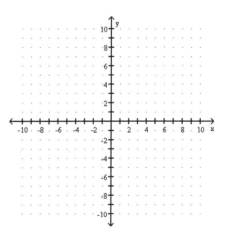

Chapter 9 Vocabulary

Vocabulary Word	Definition	Example
Square Root Property	$x^2 = a$ then $x = \pm\sqrt{a}$	$x^2 = 9$ $x = \pm\sqrt{9}$ $x = \pm 3$
Completing the Square	1. If leading coefficient is not 1, divide both sides by that coefficient. 2. Isolate all terms with the variable to one side. 3. Take $\frac{1}{2}b$, square it. Add that result to both sides. 4. Factor the perfect square trinomial. 5. Follow the square root property to solve.	$x^2 + 4x - 12 = 0$ $x^2 + 4x = 12$ $x^2 + 4x + 4 = 12 + 4$ $(x+2)^2 = 16$ $x + 2 = \pm 4$ $x = \pm 4 - 2$ $x = \{2, -6\}$
Quadratic Formula	$ax^2 + bx + c = 0$ $x = \dfrac{-b \pm \sqrt{b^2 - 4ac}}{2a}$	$2x^2 + x - 3$ $x = \dfrac{-1 \pm \sqrt{1^2 - 4(2)(-3)}}{2(2)} = \dfrac{-1 \pm \sqrt{1+24}}{4} =$ $\dfrac{-1 \pm \sqrt{25}}{4} = \dfrac{-1 \pm 5}{4} = \dfrac{4}{4}$ and $-\dfrac{6}{4}$ $x = \left\{1, -\dfrac{3}{2}\right\}$
Imaginary unit	$i,$ $i^2 = -1$ $i = \sqrt{-1}$	$\sqrt{-25} = i\sqrt{25} = 5i$
Complex number	$a + bi$ $a - bi$	$3 + 5i$ is a complex number
Complex conjugates	The conjugate of $a + bi$ is $a - bi$	$4 - 3i$ and $4 + 3i$ are complex conjugates

● **Practice Test A**

Solve using the square root property.

1. $5x^2 = 500$

1. _____

2. $(x+2)^2 = 6$

2. _____

Solve by completing the square.

3. $x^2 - 24x - 10 = 0$

3. _____

4. $3x^2 - 12x = 12$

4. _____

● Solve by using the quadratic formula

5. $x^2 = 3x - 16$

5. _____

6. $4x^2 - 20x + 5 = 0$

6. _____

Perform the indicated operations. Write the resulting complex number in standard form.

7. $(11 - 2i)(5 - 3i)$

7. _____

8. $\sqrt{-72}$

8. _____

●

9. $\dfrac{5-i}{3+2i}$

9. _____

10. $(14-5i)-(-5-7i)$

10. _____

Solve using the most appropriate method.

11. $x^2 - 18x + 20 = 0$

11. _____

12. $4x^2 = -16$

12. _____

13. $-x^2 - 5x - 7 = 0$

13. _____

Graph each quadratic equation.

14. $y = x^2 + 5x + 6$

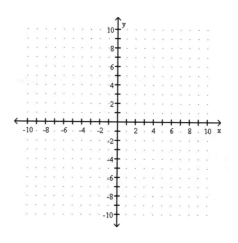

15. $y = 2x^2 - 4x + 10$

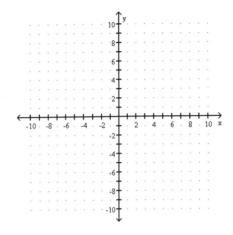

16. $y = -\dfrac{1}{6}x^2$

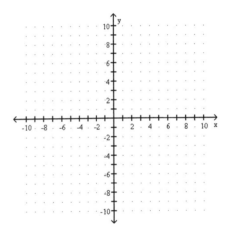

Use the vertex formula to find the coordinate of the vertex for the given quadratic equation.

17. $y = -x^2 - 3x + 4$

17. _____

18. $y = 4x^2 - 5x + 6$

18. _____

Solve.

19. John plans to fence in a garden for his vegetables. The formula he uses to figure out the area of the garden is $A = w(24 - w)$ to express the area in terms of the width of the garden. What is the maximum width that he can enclose?

19. _____

20. Find two real numbers whose sum is 11 and whose product is 18.

20. _____

Practice Test B

Solve using the square root property.

1. $x^2 = 3600$

    a. 360   b. 600

    c. 60   d. 1600

1. _____

2. $(x-7)^2 = 9$

    a. 9 and 7   b. 3 and -4

    c. 10 and 4   d. 2 and 7

2. _____

Solve by completing the square.

3. $x^2 + 12x + 9 = 0$

    a. 9 and 1   b. $\pm3\sqrt{3} + 2$

    c. $\pm3\sqrt{3} - 6$ d. 4 and 5

3. _____

4. $-2x^2 + 4x = 8$

    a. 4 and -2   b. 2 and 4

    c. 8 and -1   d. no real solution

4. _____

Solve by using the quadratic formula

5. $x^2 + 4x - 10 = 0$

    a. $\pm\sqrt{14} - 2$ b. 5 and 2

    c. 10 and 1   d. $\pm\sqrt{10} - 4$

5. _____

6. $3x^2 - 12x + 16 = 0$

6. _____

    a.-2, -8   b. $\pm\sqrt{3} + 7$

    c. $\pm\sqrt{15} + 12$ d. no real solution

Perform the indicated operations. Write the resulting complex number in standard form.

7. $-4i(2 + 7i)$

7. _____

    a. $3 - 6i$ b. $28 - 8i$

    c.- $8 + 4i$ d. $28 - 2i$

8. $\dfrac{8 - i}{2 + 4i}$

8. _____

    a. $-i$ b. $\dfrac{4}{3} + 8i$

    c. $4 - \frac{1}{4}i$ d. $\dfrac{3}{5} - \dfrac{17}{10}i$

9. $-5\sqrt{-20}$

9. _____

    a. $-10\sqrt{5}i$ b. $5\sqrt{20}i$

    c. $-10\sqrt{2}i$ d. $-5\sqrt{5}i$

Martin-Gay *Beginning Algebra, Fifth Edition*

10. $(6-i)-(16+7i)$

10. _____

    a. -10 + 6i b. -10 – 8i

    c. -22 + 8i d. 22 – 6i

Solve using the most appropriate method.

11. $-3x^2-2x-5=0$

11. _____

    a. 5 and - 1   b. $-\frac{5}{3}$ and 1

    c. $\frac{2}{3}$ and 1   d. $\frac{3}{5}$ and 5

12. $7x^2=49$

12. _____

    a. 7 and - 7   b. 1 and -1

    c. 0   d. 49 and - 49

13. $p^2+6p+4=0$

13. _____

    a. $\pm\sqrt{2}+5$ b. $\pm\sqrt{5}-3$

    c. $\pm\sqrt{3}+5$ d. $\pm\sqrt{2}-3$

Graph each quadratic equation.

14. $y = x^2 + 6x + 9$

14. _____

a.

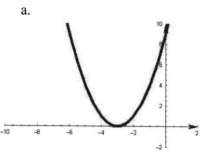

b.

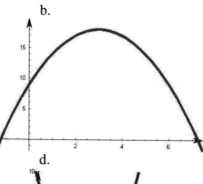

c.

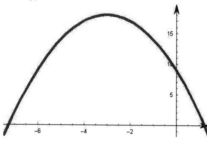

d.

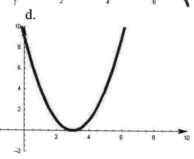

15. $y = 8x^2 + 2x - 1$

15. _____

a.

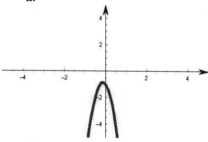

b.

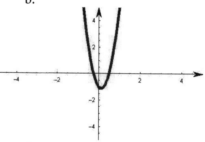

c.

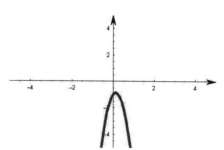

d.

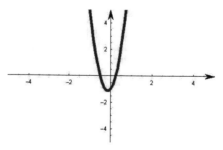

16. $y = \frac{2}{3}x^2$

16. _____

a.

b.

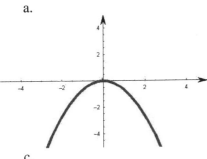
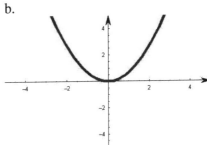

c.

d.

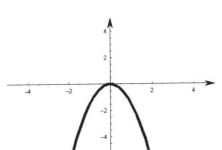
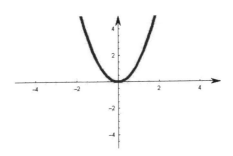

Use the vertex formula to find the coordinate of the vertex for the given quadratic equation.

17. $y = x^2 - 2x - 8$

17. _____

   a. (0, 4)

   b. (-1, -9)

   c.(1, -9)

   d. (4, 0)

18. $y = \frac{1}{2}x^2 + 6$

18. _____

   a. (0, 6)

   b. (12, 78)

   c. (-4, 14)

   d. (4, 14)

Solve.

19. The maximum height of a kicked football follows the graph of the quadratic equation
$h(t) = -16t^2 + 32t$ where t is time in seconds. What is the maximum height the football with reach?

19. _____

    a. 1 second   b. 0 seconds

    c. 3 seconds   d. 2 seconds

20. A boxing ring is in the shape of a square. Each side of the boxing ring measures 20 ft. How far away from each other are two boxers, if they are in opposite corners?

20. _____

    a. $20\sqrt{2}$ feet   b. 42 feet

    c. 29 feet   d. 20 feet

ANSWERS

Chapter 1

Section 1.2

1. Zero

3. Irrational numbers

5. Inequality symbols

7. Rational numbers

9. Integers

11. >

13. >

15. F

17. T

19. $-8 > -15$

21. $5 \geq 4$

23. integer, real, rational

25. rational real

27. -6 pounds

29. -15 degrees

31. 5

33. 8.9

35. =

37. >

39. T

41. T

43. July

Section 1.3

1. Equivalent
3. Simplified

5. Reciprocals

7. $2 \cdot 2 \cdot 2 \cdot 2$

9. $2 \cdot 2 \cdot 3 \cdot 5$

11. $\frac{1}{2}$

13. $\frac{3}{7}$

15. $\frac{8}{7}$ or $1\frac{1}{7}$

17. 1

19. $\frac{1}{4}$

21. $\frac{7}{24}$

23. $\frac{9x}{30x}$

25. $3\frac{1}{11}$

27. No

Section 1.4

1. Subtract
3. Multiply
5. Solution
7. Base; equivalent
9. 27
11. 0.0000000032
13. 40

15. $\frac{41}{50}$

17. $\frac{5}{6}$

19. $\frac{43}{8}$ or $5\frac{3}{8}$

21. No

23. $\frac{2}{3}x + 6$

25. $3(x+8)$
27. $5 - x \geq -7$
29. Answers may vary.

Section 1.5

1. Opposites
3. $-n$

5. 968

7. – 165

9. – 55

11. 2.7

13. 12° F

15. – 8

17. – 17

19. – 5

21. A. x plus negative y

    B. The opposite of x plus negative y.

    C. the opposite of x minus y.

Section 1.6

1. Supplementary

3. Complementary

5. x – 23

7. x – 23

9. 19

11. $-\frac{41}{35}$ or $-1\frac{6}{35}$

13. – 27

15. – 26

17. 0

19. 9

21. 113

23. – 7

25. 124°

27. A. T

    B. F

    C. F

    D. F

Section 1.7

1. Positive

3. Reciprocal

5. Negative

7. 0

9. $-\frac{8}{27}$

11. - 120

13. 2

15. – 32

17. – 64

19. $-\frac{6}{5}$ or $-1\frac{1}{5}$

21. – 6

23. $\frac{20}{27}$

25. – 44

27. 54

29. 1

Section 1.8

1. Commutative property of addition

3. Additive inverses

5. Associative property of addition

7. Identity element of multiplication

9. 16 + x

11. 3k + 5j

13. 10(cs)

15. 4 + (9+b)

17. $\left(-\frac{8}{13}\cdot\frac{39}{16}\right)\cdot s$

19. –r + 3 + 7p

21. -23x + 70

23. 11(x+y)

25. $\frac{1}{2}\left(\frac{1}{2}t+5\right)$

27. Answers may vary.

Practice Test A

1. 2x + 8 < 15

2. $21-\frac{1}{2}x$

3. 162

4. $\frac{68}{77}$

5. – 24

6. $16\frac{1}{4}$ or $\frac{65}{4}$

7. 0

8. $\frac{10}{9}$ or $1\frac{1}{9}$

9. – 11

10. – 83

11. – 8

12. <

13. <

14. =

15. A. 1, $\sqrt{25}$

    B. 1, 0, $\sqrt{25}$

    C. 12,-4, 1, 0, $\sqrt{25}$

    D. 12,-4, 1, 0, $\sqrt{25}$, 11,75, $\frac{5}{9}$

    E. Π

    F. All of them

16. 63

17. Identity Element of Addition

18. Associative property of addition

19. Identity element of multiplication

20. Distributive property

21. $-\frac{1}{15}$

22. $\frac{5}{7}$

23. – 28 pts.

24. The first contractor

25. Yes

Practice Test B

1. C

2. A

3. B

4. D

5. C

6. A

7. B

8. B

9. C

10. B

11. B

12. C

13. B

14. B

15. A

16. B

17. C

18. A

19. C

20. A

21. B

22. B

23. B

24. B

25. B

Chapter 2

Section 2.1

1. Exponent

3. Numerical coefficient

5.

$3x^4$	3	x	4
$\frac{1}{5}x^3$	$\frac{1}{5}$	x	3
-x	-1	x	1
7	7		0

7. -12y - 6s

9. – 7.5x + 6.2xy

11. -3y + 2x – 4z

13. 4.3y + 5.2

15. 8z – 27y

17. -3x + 18

19. It would double.

Section 2.2

1. Solving

3. Equation, expression

5. Addition

7. – 13.9

9. -0.7

11. $-\frac{33}{16}$

13. 13

15. 20 – P

17. 90 – w

19. $200 - m$

21. 6x + 16y

Section 2.3

1. Multiplication

3. Multiplication

5. 56

7. 0

9. 25

11. -1

13. $-\dfrac{13}{7}$

15. x + 27

17. 4x + 4

19. Answers will vary.

Section 2.4

1. $-\dfrac{36}{17}$

3. – 11

5. 9

7. $\dfrac{3}{4}$

9. $-\dfrac{18}{7}$

11. 8

13. 0

15. No solution

17. 0

19. 10

Section 2.5

1. 4

3. – 17

5. – 9

7. Chicago = 46
   New England = 10

9. 30 ft.

11. 31 m, 18 m

13. $14\dfrac{3}{5}$ in, $16\dfrac{3}{5}$ in, $43\dfrac{3}{5}$ in

15. 58°, 60°, 62°

17. 8, 10, 12

19. 4, 6, 8

Section 2.6

1. 32

3. 4

5. 25°

7. $43\dfrac{1}{3}$ mph

9. approx. 12 years

11. 78 ft by 36 ft

13. $\dfrac{2A}{h} = b$

15. $\dfrac{S - 2\pi r^2}{2\pi r} = h$

17. $\dfrac{3V}{\pi r^2} = h$

19. Answers may vary

Section 2.7

1. 60

3. 90

5. 55%

7. $78

9. 12.5%

11. $46.58

13. 11.11%

15. $53,333.33

17. 12 lbs

19. 150 L

21. Yes.

Section 2.8

1. 6.5 hrs

3. 17 hours

5. 82.5 miles

7. 15 dimes 60 nickels

9. 41 DVD players
   123 MP3 players

11. 3 times

13. $1220 at 5%
    $2580 at 7%

15. $211.11 at 10%
    $5888.89 at 8%

17. $4800 at 9%
    $3600 at 12%

19. 8.57 L

Section 2.9

1. False

3. Linear Inequality in One Variable.

5. x $\leq$ - 1

7. $(-\infty, -1]$

9. $[6, \infty)$

11. $[-5, \infty)$

13. $(-\infty, 7)$

15. $(-7, \infty)$

17. $[-6, 15)$

19. $(-8, 46)$

Martin-Gay *Beginning Algebra, Fifth Edition*

21. $x > -2$
23. $x < 150$

Practice Test A

1. $-4t + 18$
2. $x - 30$
3. $-3.32p - 6.6$
4. $\dfrac{17}{28}x - \dfrac{18}{7}$
5. -8
6. -20
7. $24\!/\!5$
8. -11
9. No Solution
10. -11
11. 15
12. 430
13. All real numbers
14. 33
15. 20
16. 9, 11, 13
17. 61.67 kg.
18. 6.5 hrs.
19. 160 ft.
20. $\dfrac{S - P}{\text{Pr}} = t$
21. $m = \dfrac{y - b}{x}$
22. $\left(4\dfrac{1}{2}, \infty\right)$
23. $[4, \infty)$
24. $[6, 12]$
25. between 2 and 5 ft.

Practice Test B

1. C
2. D
3. A
4. B
5. A
6. B
7. D
8. A
9. C
10. B
11. D

12. C
13. C
14. C
15. B
16. C
17. C
18. B
19. A
20. A
21. D
22. C
23. B
24. B
25. D

Chapter 3
Section 3.1

1. y-axis, x-axis
3. Origin, (0, 0)
5. x-coordinate, y-coordinate
7. 1997 to 1998
9. Yes. Answers vary
11. 250
13.

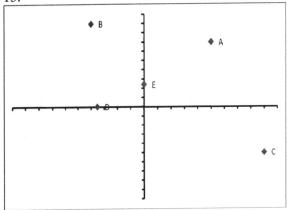

15. yes, no, no

17. (-4, -2) (4, 0)
19.

0	0
2	-8
-3	12

21. a. (-1.6, -2.2)
   b. 24.8
   c. 36.48

Section 3.2

1. yes
3. yes
5.

0	0
5	0
2	3

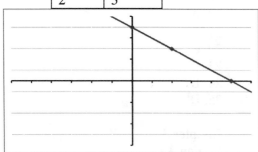

7.

0	-2
1	-5
2	-8

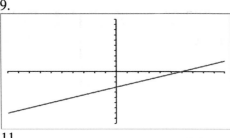

9.

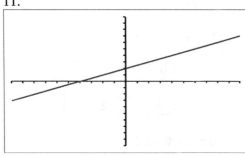

11.

13.

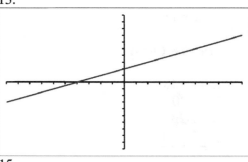

15.

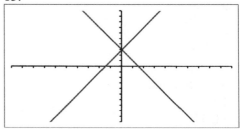

Both graphs pass through (0, 3)
17. $700

Section 3.3

1. Horizontal
3. Standard
5. Vertical
7. (0, 6) (-6, 0)
9. (0, -4) (6,0)

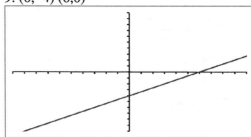

11. (0, 3) (-6, 0)

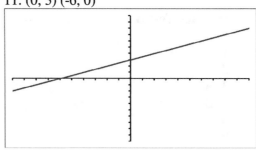

Martin-Gay *Beginning Algebra, Fifth Edition*

13. (0,0)

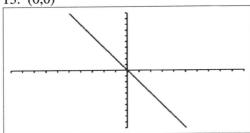

15.

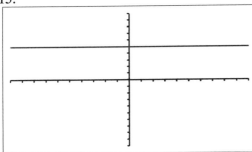

17. Graph is the y-axis.

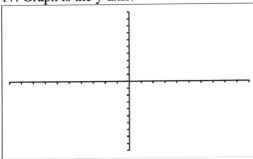

19. No. answers vary

Section 3.4

1. m, b
3. Undefined
5. Parallel
7. m = -1
9. m = undefined
11. $m = \frac{2}{3}$
13. $m = \frac{4}{5}$
15. m = 0
17. perpindicular
19. $\frac{1}{2}$
21. a. (0, 31) (22, 54)

   b. $y = \frac{14}{11}x + 31$

   c. ≈ 70.15%

   d. 2031

Section 3.5

1. Point slope
3. Slope-intercept
5. Vertical
7. $y = \frac{2}{3}x - 2$
9.

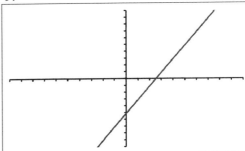

11.

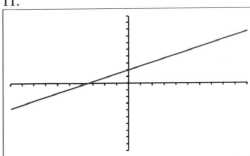

13. 8x + y = -13
15. 3x + y = 1
17. $\frac{7}{4} + y = -5$
19. x = 24
21. x = -7
23. (-8, -2); $y = \frac{2}{3}x + \frac{10}{3}$

Section 3.6

1. Vertical
3. Function
5. Range
7. {2, 8, -7, 1}
9. yes
11. yes
13. no
15. yes
17. yes
19. (-1, 10) (2, 16) (0,6)
21. Domain $(-\infty, \infty)$ Range $(-\infty, \infty)$

23. $\left(-\infty,\dfrac{3}{2}\right)\cup\left(\dfrac{3}{2},\infty\right)$

25a. 166.38 cm

   b. 148.25 cm

Practice Test A

1. (-3, 13) (0, 4) (7, -17)
2. III
3. no
4. no
5.

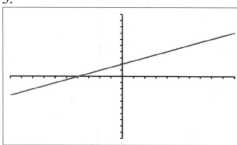

6.

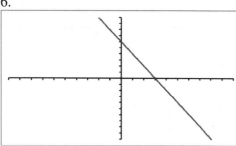

7. (0, -5) (-3, 0)
8. (0, 3) (-2, 0)

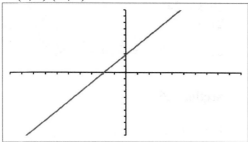

9. (0, 0)

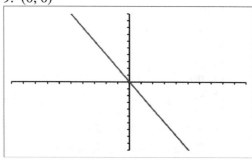

10.

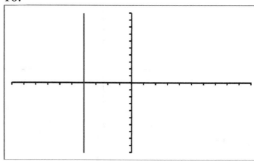

11. $m = -\dfrac{3}{2}$

12. $m = -\dfrac{1}{3}$

13. $m = \dfrac{3}{7}$

14. neither

15. $m = -\dfrac{5}{9}$ $b = -\dfrac{4}{3}$

16. $y = \dfrac{4}{7}x - 6$

17.

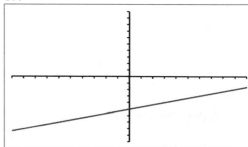

18.

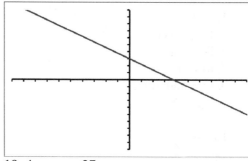

19. 4x + y = -27
20. 2x − 5y = -26
21. x = -14
22. y = 24
23. yes
24. yes
25. (-4, 37) (0, 9) (3, 9)

Practice Test B

1. D
2. D
3. A
4. A
5. C
6. B
7. C
8. D
9. B
10. D
11. A
12. B
13. B
14. D
15. C
16. B
17. B
18. A
19. C
20. A
21. A
22. B
23. C
24. A

Chapter 4
Section 4.1

1. Independent
3. Consistent
5. Inconsistent
7a. Yes
7b. No
9a. No
9b. Yes
11. (-1, -2)
13. Infinite Solutions
15. Infinite lines; Identical lines
17. 1 solution; Intersecting lines

Section 4.2

1. (-3, 9)
3. All Real Numbers
5. (-2, 2)
7. (-2, -1)
9. Infinite Solutions

11. $\left(-\frac{1}{3}, \frac{2}{3}\right)$
13. No solution
15. No solution
17. $\left(\frac{1}{4}, \frac{1}{5}\right)$
19. $\left(\frac{28}{13}, -\frac{94}{13}\right)$

Section 4.3

1. (2, -3)
3. $\left(-\frac{16}{25}, -\frac{2}{5}\right)$
5. All Real Numbers
7. No Solution
9. All Real Numbers
11. (56, 65)
13. $\left(\frac{10}{17}, \frac{14}{17}\right)$
15. (-1, 3)
17. $\left(\frac{11}{71}, \frac{14}{71}\right)$
19. (1, 4)

Section 4.4

1. (50, 33)
3. Length = 28 ft. Width = 13 ft.
5. 6 and 9
7. 100 nonstudents and 250 students
9. adult = $29 and children = $18
11. 2 hours
13. 12 liters of 20% and 6 liters of 50%
15. $11750 at 7% and $18250 at 9%
17. 28 by 15

Section 4.5

1. Boundary Line
3. False
5. Solution
7.

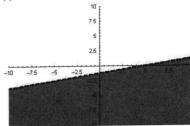

9.

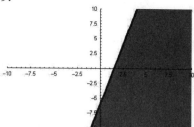

11.

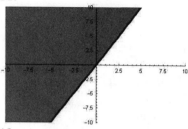

13.

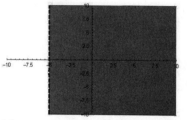

15.

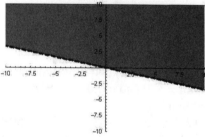

17.

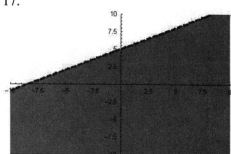

Section 4.6

1.

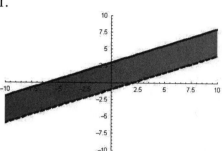

3.

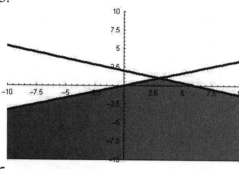

5.

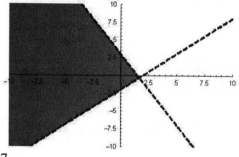

7.

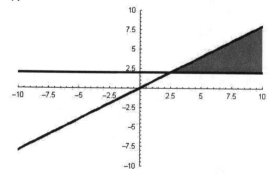

Martin-Gay *Beginning Algebra, Fifth Edition*

9.

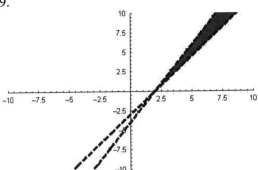

4. (-2, -1)
5. (41, 32)
6. (-2, -6)
7. $\left(1, -\frac{3}{2}\right)$
8. $\left(\frac{1}{4}, \frac{11}{8}\right)$
9. $\left(-\frac{4}{3}, 4\right)$
10. (-2, 2)
11. All Real Numbers
12.

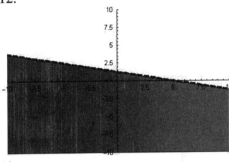

11.

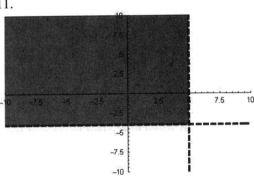

13.

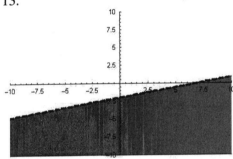

13.

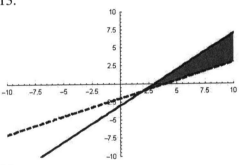

14.

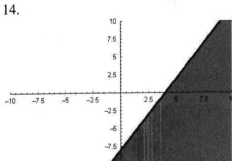

15.

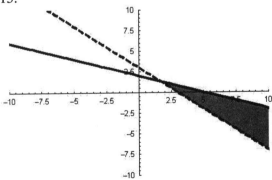

15.

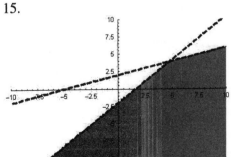

Practice Test A

1. Yes
2. No
3. (-5, -3)

16.

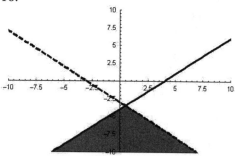

17. 8 and -1
18. 5 liters of 25% and 10 liters of the 40%
19. 28 by 18
20. $1800 at 4% and $3600 at 10%

Practice Test B

1. A
2. A
3. B
4. A
5. B
6. A
7. C
8. A
9. C
10. B
11. A
12. C
13. A
14. D
15. B
16. C
17. C
18. B
19. A
20. A

Chapter 5
Section 5.1

1. Base, exponent
3. Multiply
5. 1
7. 0.000027
9. $2^{11}x^{18}$
11. $x^{19}y^{6}$

13. $(-7)^{48}$
15. $a^{24}b^{30}$
17. 0.0008 y^{48}
19. $\dfrac{4x^{2}z^{4}}{y^{10}}$
21. 1
23. $\dfrac{1}{9}$
25. $16mn^{3}$
27. It will be 64 times larger

Section 5.2

1. Binomial
3. True
5. Coefficient
7. 6th degree; Trinomial
9. 8th degree; None of these
11. 0
13. 4
15. 15ab - 6a - 8b
17. – 10x – 13
19. $x^{3}-7x^{2}+4x+3$
21. $-2m^{2}+12m-10$
23. 112 ft

Section 5.3

1. $8x^{5}-6x^{3}$
3. $-5x^{3}y+7x^{2}y^{2}-3y^{4}$
5. $\dfrac{1}{10}x^{2}-\dfrac{23}{10}x+\dfrac{1}{12}$
7. $ab-10a+7b-70$
9. $9x^{4}+6x^{2}+1$
11. $x^{3}-6x^{2}+10x-4$
13. $a^{2}-3a-4$
15. $10x^{3}+22x^{2}-x-1$
17. $6x^{2}+16x+2$

Section 5.4

1. $x^{2}-x-42$
3. $33y^{2}-80xy+48x^{2}$
5. $16m^{2}+6mp-p^{2}$

7. $x^2 - \frac{1}{2}x + \frac{1}{16}$

9. $49s^2 - 42st + 9t^2$

11. $x^2 - 25$

13. $81x^2 - y^2$

15. $81x^2 - 169$

17. $625x^4 - 1000x^3 + 600x^2 - 160x + 16$

Section 5.5

1. Scientific notation

3. x^7

5. $\frac{1}{512}$

7. $\frac{-4}{x^6}$

9. $\frac{-8}{9}$

11. $\frac{1}{x^4}$

13. $-16b^{11}$

15. 1.67×10^{-6}

17. 0.0000003478

19. 200,000

Section 5.6

1. $-3x^2 + \frac{5}{2}x - 2$

3. $-3x^2 + x - \frac{4}{x^3}$

5. $x + 1$

7. $4m^2 + 6m - 8 + \frac{10}{2m+3}$

9. $2b^2 + b + 2 - \frac{12}{b+4}$

11. m

Practice Test A

1. $\frac{1}{512}$

2. -64

3. $\frac{r^{15}s^{11}}{2}$

4. $-108m^{15}n^7$

5. 1.372×10^{11}

6. 0.00000004284

7. 5

8. $x^2y + 5xy^2 + 2y^2$

9. $9a^2 + 6a - 2$

10. $-10x^2 + 8x + 9$

11. $-2c^2 + 2c - 6$

12. $2a^2 - ab - 3b^2$

13. $-4x^4 + 5x^3 - 2x^2 + 8x$

14. $4m^3 - 16m^2 - 6mn + 2m^2n + n^2$

15. $x^2 - 169$

16. $y^2 - 14y + 49$

17. $49x^2 - \frac{1}{4}$

18. $c^3 + 12c^2 + 48c + 64$

19. $2x^2 - 5x$

20. -252 ft

21. $2x^2 - 4x - 3$

22. $\frac{3}{4}m - n - \frac{2}{n}$

23. $2x + 7$

24. $x^2 + 5x + 25$

25. $3x^2 + 2x - 4 + \frac{4}{2x-1}$

Practice Test B

1. A
2. D
3. D
4. D
5. B
6. C
7. A
8. A
9. C
10. D
11. B
12. A
13. C
14. D
15. A
16. C
17. B
18. A

19. D
20. D
21. C
22. C
23. B
24. D
25. A

Chapter 6
Section 6.1

1. Factors
3. Greatest common factor
5. 3
7. 1
9. $4y^3$
11. $3a^4b^3$
13. $3\left(3x^2 - 6x + 2\right)$
15. $\left(x^2 + 2\right)(y + 3)$
17. $(b - 4)(a + 5b)$
19. $3(8a - 9)(7b + 4)$

Section 6.2

1. $(x-6)(x+3)$
3. $(x + 8)(x + 4)$
5. $(x + 1)(x - 3)$
7. $(x + 4y)(x - 5y)$
9. prime
11. $4y\left(4x^2 - 4x - 45\right)$
13. $\dfrac{1}{2}(x + 4)(x - 8)$
15. $b = 7$

Section 6.3

1. $(3x + 2)(2x - 1)$
3. prime
5. $(3x - 5)(9x - 4)$
7. $x(4x + 3)(x - 3)$
9. $7(8m - 9)(m + 1)$
11. $\left(4x - 5\right)^2$
13. $4\left(2x - 3\right)^2$
15. $\left(6x - \dfrac{1}{2}\right)^2$

17. $\dfrac{1}{12}\left(3x - 2\right)\left(x - 2\right)$

Section 6.4

1. $(x + 2)(x + 3)$
3. $(a - 4)(2a - 5)$
4. $(2x - 1)(6x - 7)$
5. $12(3x + 1)(x - 2)$
7. prime
9. $\dfrac{1}{4}(2x - 1)(x - 6)$

Section 6.5

1. Difference of squares
3. Perfect square trinomial
5. $(x + 7)(x - 7)$
7. $(11m + 10n)(11m - 10n)$
9. prime
11. $\left(\dfrac{5}{6}t + \dfrac{2}{3}\right)\left(\dfrac{5}{6}t - \dfrac{2}{3}\right)$
13. $(y + 3)(y^2 - 3y + 9)$
15. prime
17. $\left(x^2 + y^3\right)\left(x^2 - y^3\right)$
19. $(x - 4)\left(x^2 + x + 7\right)$

Section 6.6

1. $\{-3, 4\}$
3. $\left\{\dfrac{7}{3}, \dfrac{1}{2}\right\}$
5. $\{-7\}$
7. $\{0, 1, -1\}$
9. $\{0, -3, 8\}$
11. $\{0, 3, -1\}$
13. $\left\{\dfrac{1}{2}, -6\right\}$
15. $\{8, 0\}$

Section 6.7

1. 14, 19, 52
3. 12, 14

5. 11, 13

7. L = 8 in, W = 6 in.

9. $\dfrac{9}{4}$ and $\dfrac{1}{4}$

11. 12, 16

Practice Test A

1. $(x - 7)(x + 4)$

2. $(2x + 5)(3x - 2)$

3. $(2x+1)\left(4x^2 - 2x + 1\right)$

4. $3(x - 9)(x - 1)$

5. $\left(x^2 + 4\right)(x + 2)(x - 2)$

6. $- (x- 4)(4x - 7)$

7. $(x-2)\left(3x^2 - 5\right)$

8. $(2x + 13)(2x - 13)$

9. $(x + 6)(x + 5)$

10. prime

11. $\{4, 9\}$

12. $\{0, -1, -6\}$

13. $\left\{\dfrac{1}{5}, -\dfrac{3}{2}\right\}$

14. $\left\{-\dfrac{2}{5}, -\dfrac{2}{3}\right\}$

15. $\left\{-\dfrac{4}{5}\right\}$

16. $\left\{1, -\dfrac{1}{6}\right\}$

17. $\left\{\dfrac{1}{11}, -2\right\}$

18. $\left\{\dfrac{1}{5}\right\}$

19. $\{0, -5\}$

20. 4.5 seconds

Practice Test B

1. A
2. B
3. C
4. A
5. B
6. D
7. B
8. A
9. A

10. A
11. C
12. C
13. B
14. D
15. B
16. B
17. C
18. D
19. B
20. C

Chapter 7
Section 7.1

1. $-\dfrac{2}{3}$

3. $\dfrac{11}{12}$

5. none

7. none

9. $\dfrac{11}{x+1}$

11. $\dfrac{x^2}{x-2}$

13. $\dfrac{-x-11}{x-4}, \dfrac{-(x+11)}{x-4}, \dfrac{x+11}{-x+4} \cdot \dfrac{x+11}{-(x-4)}$

15. $\dfrac{-x+5}{9-x}, \dfrac{-(x-5)}{9-x}, \dfrac{x-5}{-(9-x)}, \dfrac{x-5}{x-9}$

Section 7.2

1. $\dfrac{5w}{9x^2}$

3. $\dfrac{4}{x+3}$

5. $\dfrac{1}{z+1}$

7. $\dfrac{7y^2}{2x^2}$

9. $\dfrac{x-4}{x-3}$

11. $\dfrac{(x-3)(x-1)}{x+1}$

13. 1

15. $\dfrac{(x+3)^2 (x-1)}{(x+4)^2 (x-2)}$

Section 7.3

1. $\dfrac{3x-10}{x+2}$

3. $\dfrac{-6x-4}{x+3}$

5. $\dfrac{-5x+8}{x+1}$

7. $(x + 1)(x - 3)$

9. $(x + 4)(x - 4)(x - 4)$

11. $(x + 1)(x - 1)(x^2 - x + 1)$

13. $\dfrac{3x(x-3)}{x^3 - 5x^2 + 6x}$

15. $\dfrac{(x-1)(x+2)}{x^3 + 6x^2 - x - 6}$

Section 7.4

1. $\dfrac{7x-4}{x^2 - 1}$

3. $\dfrac{-y-4}{y+3}$

5. $\dfrac{2\left(x^2 + 7x + 6\right)}{(x+3)(x+3)(x+5)}$

7. $\dfrac{-2(y+1)}{(y-4)(y+3)(y+1)}$

9. $\dfrac{2x^2 + 3x - 1}{(x-8)(x-1)(x+1)}$

11. $\dfrac{2(x^2 - x - 23)}{(x+1)(x-5)(x-6)}$

13. $\dfrac{x^2 + 15x - 22}{(x-6)(x-2)(x+2)}$

Section 7.5

1. $\left\{ \dfrac{7}{5} \right\}$

3. $\{8, -2\}$

5. No solution

7. Mo solution

9. $\left\{ -\dfrac{9}{13} \right\}$

11. $p = \dfrac{qr}{q-r}$

13. $B = \dfrac{2U + TE}{T}$

15. Answers may vary

Section 7.6

1. 33

2. – 3

3. 7

4. -4

5. 60 women

6. 915 dentists

7. 4560 m

8. $\dfrac{11}{3}$

9. 4

10. 4 and 6

11. 3.5 minutes

12. 2 hours

13. 3.75 hours

14. motorcycle is 60 mph and the car is 70 mph

15. Lucy walks at 1 mph, Michelle walks at 2 mph

16. 150 mph

Section 7.7

1. $y = kn$

3. 40

5. $y = \dfrac{k}{t}$

7. 24

9. $x = ky^2$

11. $\dfrac{12}{27}$

13. 150 pounds

15. 10 cubic centimeters

Section 7.8

1. $\dfrac{3m - 2n}{7m + 1}$

3. $\dfrac{2x(x-5)}{x^2+10}$

5. $\dfrac{-x^2+6x-4}{3x-2}$

7. $\dfrac{9x-23}{3x-7}$

9. $\dfrac{x+y}{x-y}$

11. $\dfrac{-25x^2+100}{4x^2-100}$

Practice Test A

1. $\{3, 2\}$
2. $\{2, -2\}$
3. $\{0, 6\}$

4. $\dfrac{x^2+2x-7}{x+1}$

5. $\dfrac{2x}{(x+4)(x-5)}$

6. $\dfrac{(x-2)^2}{(x+2)(2x+1)}$

7. $\dfrac{2x}{(x-5)(x+3)}$

8. -1

9. $\dfrac{(x-4)(x+1)}{3x}$

10. $\dfrac{5}{8(x+2)}$

11. $\dfrac{x(2x-1)}{4-3x^2}$

12. $\dfrac{4x^2-x+3}{5x+9}$

13. $-\dfrac{3}{7}$

14. $\{3, 2\}$
15. 8

16. $\dfrac{m}{1-mt}$

17. $B = \dfrac{bh-2A}{-h}$

18. 18 inches
19. 1.2 hours
20. 15

Practice Test B

1. B
2. A
3. C
4. D
5. A
6. B
7. A
8. C
9. B
10. C
11. C
12. D
13. D
14. C
15. B
16. D
17. B
18. D
19. A
20. B

Chapter 8
Section 8.1

1. Radicand, index
3. Perfect
5. -11

7. $\dfrac{2}{5}$

9. 4

11. $\dfrac{1}{5}$

13. $\dfrac{6}{7}$

15. -8

17. $\dfrac{1}{2}$

19. 15.524
21. 6.633
23. $9y^3z^4$
25. $x^4y^5z^2$
27. $|x+5|$

Section 8.2

1. $6\sqrt{5}$

3. $-12\sqrt{2}$

5. $10\sqrt{2}$

7. $\dfrac{-3\sqrt{5}}{8}$

9. $\dfrac{-\sqrt{30}}{6}$

11. $4t^3\sqrt{2}$

13. $\dfrac{5m\sqrt{10m}}{3n^3}$

15. $2\sqrt[5]{5}$

17. not a real number

19. 8.15 seconds

Section 8.3

1. $-3\sqrt{2}$

3. $-3\sqrt[3]{6}+4\sqrt{6}$

5. $4\sqrt{10}$

7. $-3x\sqrt{2}$

9. $\dfrac{3\sqrt{3}}{8}$

11. $6u^2t\sqrt[4]{5t}$

Section 8.4

1. $8\sqrt{10}$

3. $\sqrt{6}-\sqrt{15}$

5. $7y\sqrt{3}+\sqrt{42xy}-3\sqrt{7xy}-3x\sqrt{2}$

7. $2+2\sqrt{10x}+5x$

9. $5y$

11. $-6x\sqrt{3}$

13. $\dfrac{\sqrt{18a}}{6a}$

15. $\dfrac{2\sqrt{7mn}}{7n}$

17. $\dfrac{\sqrt[3]{3}}{3}$

19. $\dfrac{21+7\sqrt{x}}{9-x}$

21. $\dfrac{24+6\sqrt{3}-4\sqrt{5}-\sqrt{15}}{13}$

23. $\dfrac{5}{4\sqrt{5}}$

Section 8.5

1. 40

3. no solution

5. -1

7. 6

9. no solution

11. no solution

13. 2

15. 29

Section 8.6

1. $2\sqrt{5}$

3. $\sqrt{22}$

5. $2\sqrt{10}$

7. $\sqrt{61}$

9. $\dfrac{\sqrt{101}}{4}$

11. 20.6 feet

13. $6\sqrt{5}$

15. $\dfrac{\sqrt{985}}{10}$

Section 8.7

1. 6

3. 0.5

5. -9

7. 4

9. 16

11. $-\dfrac{1}{2}$

13. $\dfrac{5}{4}$

15. $2^{5/6}$

17. x^5

19. $\dfrac{1}{2^{1/3}}$

21. 2

Martin-Gay *Beginning Algebra, Fifth Edition*

ANSWERS

Practice Test A

1. $5x^2y^3\sqrt{5}$

2. $\dfrac{1}{x^4y^6z^9}$

3. $3\sqrt{x}$

4. $\sqrt{3}-\sqrt{15}$

5. $2u^3v\sqrt{5v}$

6. $11\sqrt{2}-2\sqrt{3}$

7. $-4x\sqrt{2x}+28\sqrt{2}$

8. $\dfrac{10-2\sqrt{7}}{9}$

9. $\dfrac{3\sqrt{2}-\sqrt{30}-\sqrt{6}+\sqrt{10}}{-2}$

10. $\dfrac{9\sqrt{10}}{10}$

11. no solution

12. 5

13. $37\!\big/\!4$

14. $\{2, -5\}$

15. 9 inches

16. $\sqrt{170}$

17. $2\sqrt{5}$ inches

18. $6\sqrt{2}$ feet

19. $5^{6/5}$

20. $\dfrac{1}{7^{1/4}}$

Practice Test B

1. A
2. B
3. C
4. A
5. A
6. B
7. D
8. A
9. B
10. A
11. A
12. C
13. A
14. B

15. C
16. B
17. B
18. D
19. C
20. B

Chapter 9
Section 9.1

1. 12 and -12
3. 5 and -5
5. 8 and -2
7. $\pm\sqrt{10}-2$
9. $\dfrac{\pm3\sqrt{10}}{2}$
11. $\pm2\sqrt{3}+4$
13. $12\sqrt{2}$
15. $\pm2\sqrt{3}+3$

Section 9.2

1. $x^2-2x+1; (x-1)^2$
3. $x^2+7x+\dfrac{49}{4}; \left(x-\dfrac{9}{2}\right)^2$
5. $\{-2, -6\}$
7. $\{0, -1\}$
9. no solution
11. $\dfrac{\pm\sqrt{57}+3}{3}$
13. $\dfrac{\pm\sqrt{41}+5}{8}$
15. 16

Section 9.3

1. $\dfrac{\pm\sqrt{37}-7}{6}$
3. $\dfrac{\pm\sqrt{59}+2}{5}$
5. $\dfrac{\pm\sqrt{41}-5}{2}$
7. no solution

9. 1.721, -0.387
11. 1.612, -1.612

13. no real solution
15. no real solution
17. 0.678, -4.178

Section 9.4

1. Imaginary
3. Conjugate
5. $\pm 6i$
7. $\pm\sqrt[4]{50}\,i$
9. $-2 - 4i$
11. $-9 + 2i$
13. $-10 - 5i$
15. 25
17. $-65 - 3i$
19. $8 + 9i$
21. $-5 - 6i$
23. $\pm 2\sqrt{3} - 3$
25. $\dfrac{3}{8} \pm \dfrac{\sqrt{71}}{8}\,i$
27. $\dfrac{5}{4} \pm \dfrac{\sqrt{47}}{4}\,i$

Section 9.5

1.

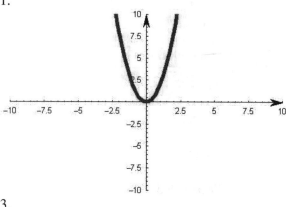

3.

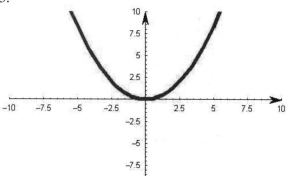

5.

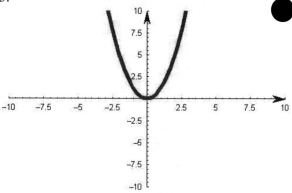

7.

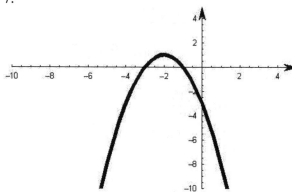

9.

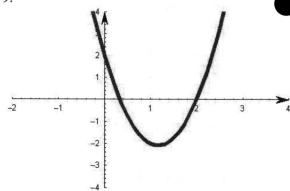

11. $(2, 2)$
13. $\left(\dfrac{5}{6}, \dfrac{-95}{12}\right)$
15. Vertex $(2, 2.4)$ No x-intercept

Practice Test A

1. 10 and -10
2. $\pm\sqrt{6} - 2$
3. $\pm\sqrt{154} + 12$
4. $\pm 2\sqrt{2} + 2$

Martin-Gay *Beginning Algebra, Fifth Edition*

5. $\dfrac{3}{2} \pm \dfrac{\sqrt{55}}{2} i$

6. $\dfrac{\pm 2\sqrt{5} + 5}{2}$

7. $49 - 43i$

8. $6\sqrt{2}i$

9. $\dfrac{17}{13} + \dfrac{7}{13} i$

10. $19 + 2i$

11. $\pm\sqrt{61} - 9$

12. $\pm 2i$

13. $-\dfrac{5}{2} \pm \dfrac{\sqrt{3}}{2} i$

14.

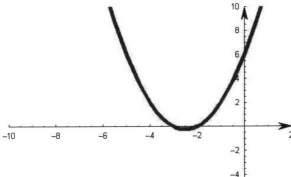

15.

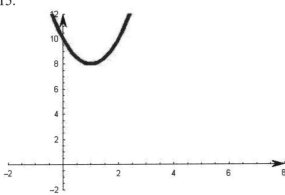

16.

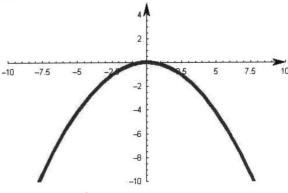

17. $\left(-\dfrac{3}{2}, \dfrac{25}{4} \right)$

18. $\left(\dfrac{5}{8}, \dfrac{71}{16} \right)$

19. 24

20. $\{9, 2\}$

Practice Test B

1. C
2. C
3. C
4. D
5. A
6. D
7. B
8. D
9. A
10. B
11. B
12. A
13. B
14. A
15. D
16. B
17. C
18. A
19. D
20. A